Tropical

Tropical Tree Crops

Lawrence K Opeke

University of Ife,
Ile-Ife, Nigeria

JOHN WILEY AND SONS

Chichester · New York · Brisbane · Toronto · Singapore

Library of Congress Cataloging in Publication Data

Opeke, Lawrence K.
Tropical tree crops.

Includes index.
1. Tree crops—Tropics. 2. Tropical crops.
3. Tree crops—Africa, West. 4. Tropical crops—
Africa, West. I. Title.
SB171.T76063 634'.0913 81–11501

ISBN 0–471–10060–9 (Cloth) AACR2
ISBN 0–471–10066–8 (Paper)

British Library Cataloguing in Publication Data:

Opeke, L. K.
Tropical tree crops
1. Tropical crops
2. Tree crops
I. Title
631. 0913 SB111

ISBN 0 471 10060 9 (Cloth)
ISBN 0 471 10066 8 (Paper)

Photosetting by Thomson Press (India) Limited, New Delhi
and printed in the United States of America,
by Vail-Ballou Press Inc., Binghamton, N. Y.

This book, *Tropical Tree Crops* is dedicated to my parents, my father Chief Lekiran Philip Opeke and my late mother Madam Kupoluyi Opeke. My father, a great cacao planter, afforded me excellent opportunity for education and it is through the inspiration I derived from his farming activity that I made a career in agriculture.

To them both,
I am very grateful.

LAWRENCE K. OPEKE

Contents

Preface

Products of tropical tree plants such as cacao, kola, coffee, tea, oil palm, coconut, rubber, citrus, and many others enjoy wide popularity among millions of people all over the world. Millions of people are engaged in their cultivation as commodity crops while others process and trade in their products. Several million hectares of fertile land in the tropics and to a less extent in the subtropics are devoted solely to their cultivation. Tropicalized greenhouses are built in many parts of the temperate zone for experimental growing of these crops. Nevertheless, these crop plants are only known through the names of their products. Many people, including those who cannot do without the traditional after dinner cup of coffee, do not appreciate the time, the technique, the ingenuity, the resources, and the pains taken by the coffee farmer to produce the raw beans from which this most cherished cup of coffee was made. In like manner, both the consumers and farmers of coffee seldom visualize the complicated factory processing which the coffee beans have to pass through before the various coffee packages appear on the supermarket shelves. Similarly, millions who cherish chocolate are not aware that the bulk of the raw material from which that bar of chocolate was made came from the cacao trees growing in West Africa or Brazil. Many do not also know that there are many other tropical plants the products of which, if developed, can still contribute substantially to the well-being of man and his livestock in terms of food, drugs, and agro-based industrial raw materials.

This book, *Tropical Tree Crops*, is intended to inform those who are interested and to fill a major gap in the body of literature on tropical agriculture. It aims at assisting both the teachers and the students of tropical agriculture and economic botany, agro-based industrialists and the general readers in their search for valuable information on the basic principles of the agronomy of tropical tree crops. I have attempted to bring together under one cover the essentials that are required for successful cultivation of tropical tree crops.

Tropical Tree Crops is written with the secondary sixth form and the university undergraduate in agriculture, forestry, economic botany, pharmacy and other biological sciences in view. It is written in simple non-technical language so that the contents can be easily understood by the non-specialists.

The book covers four distinct areas. The first six chapters discuss the general environment and the basic principles and practices in the cultivation of tree crops. These chapters highlight the effects of environmental factors on the distribution of major tropical tree crops, nursery practices and principles of propagation. Equipment and materials, planting in the field, field maintenance, regeneration and rehabilitation are also treated at considerable length. Chapters 7 to 16 give details on the origin, spread, habitat, botany, cultivation and processing of cacao, kola, coffee, citrus, latex producing plants, cashew, oil palm, coconut, mango and tea. Chapter 17 introduces the reader into the realms of the untapped wealth which is still available in the tropical forests. The chapter discusses a variety of tropical tree plants that are currently either underexploited or unexploited. The final chapter, Chapter 18, discusses the future of tropical tree crops in the international economy.

The motivation for writing this book arose from the experiences which I gained from my father's farm and my long connection with tropical tree crops. As a college scholar, I graduated B.Sc. Agric. London; Ph.D. Agric. Botany, Reading, England (specializing in Cytogenetics with Plant Breeding); and Diploma in Tropical Agriculture, Moor Plantation, Ibadan. I was honoured with a certificate of Merit in Agriculture by the Association for the Advancement of Agricultural Sciences in Africa (AAASA). I travelled extensively in many parts of the world in search of knowledge on tropical tree crops. My working life was mostly spent either in teaching or on researching about tropical tree crops. As a result of these varied experiences, I discovered that a big gap exists in the body of literature on tropical agriculture. The gap is the non-existence of a standard textbook on tropical tree crops. This book does not claim to contain all the available and relevant information on the area covered. It does, however, provide a valuable take-off point for those who are interested in tropical tree crops together with their origin, spread, ecology, botany, and principles and practices of their cultivation.

LAWRENCE K. OPEKE
May 1981

List of Figures

List of Tables

CHAPTER 1

Ecology of Tree Crops of West Africa

The ecology of the different tree crops of West Africa is closely related to the situation of the rain forest belt along the coast and the savanna zones located to the north. The common features of the general ecology are outlined below, while the specifics are discussed under each crop.

1.1 ALTITUDE AND WINDS

Most of the West African tree crops occur in the tropical lowland stretching along the coast from Guinea to Cameroun. Its elevation ranges from sea level up to an altitude of 2000 m, but most tree crops grow below 200 m. The weather is dominated by two air streams, first the northerly harmattan or dry continental tropical air which originates above the Sahara, and second the humid oceanic air brought by the south-west winds. The two air streams converge at the inter-tropical convergence (ITC), which is characterized by a thick, low cloud cover, heavy rains and violent thunderstorms. The ITC follows the seasonal changes in the position of the sun and in January it attains its southernmost position: it is then located above the coastal areas of most of West Africa, although not reaching beyond Sierra Leone and Liberia. In July, it reaches a latitude of about 20 °N.

1.2 RAINFALL

The south-west winds bring rains mainly in a zone 500–550 km wide of which the ITC forms the northern limit with a variation of about 150 km. South of this belt, less rain falls and when the belt has passed over in a northerly direction, the weather pattern shows a 'little dry season', the characteristic August break in the rains. Most of the tree crop belt has therefore a rainfall pattern with two maxima. The rainfall is spread at least over eight rainy months (with a minimum of 50 mm rainfall per month). In some areas in the eastern part of West Africa, Nigeria and Cameroun, the climate becomes of an equatorial type, also with two seasons of heavy rainfall when the sun is overhead, but with intervening months which are slightly dry. The transitional areas along the northern boundary of the equatorial climatic zone have less rain during the dry season and are denoted as semi-seasonal equatorial climates. Also, the furthest western part, Liberia and Sierra Leone, has a monsoon type of climate with a marked single maximum rainfall in summer and a short dry season.

The tree crop belt cuts right through the isohyets of average annual rainfall from 1000 to 3500 mm. The highest rainfall occurs along the south-west coast of Sierra Leone and Liberia, where the annual rainfall figures range from 2200 to 4000 mm. In an easterly direction the rainfall drops to 1400 mm in much of the forest area in Ivory Coast and Ghana. In central Ivory Coast and east-central Ghana the rainfall ranges from 1200 to 1400 mm, as it also does in most of the tree crop belt of Nigeria. However, towards the boundary of Nigeria with Cameroun and beyond rainfall increases to over 2500 mm. In the western part of the tree crop belt, the dry season may be restricted to one month with a rainfall of less than 50 mm. Further east, in Ivory Coast and Ghana, the dry season lasts from two to three months, in Nigeria from three to four months and in Cameroun one to two months.

Deviating markedly from this pattern is the rainfall in the coastal dry climate

Table 1 Average monthly rainfall (mm) at four locations within the West African tree crop belt (Adapted from Harrison Church, 1960; Wills, 1962)

	Harbel Liberia	Gaguoa Ivory Coast	Kumasi Ghana	Ibadan Nigeria
Months	6°24′N 10°25′W	6°06′N 5°55′W	6°45′N 1°40′W	7°10′N 3°50′E
January	25	32	17	10
February	42	60	58	23
March	115	144	135	88
April	155	142	141	135
May	292	186	179	148
June	407	228	230	185
July	466	93	124	158
August	505	61	73	83
September	612	174	174	175
October	365	164	199	153
November	185	100	97	45
December	72	56	30	10
TOTAL	3241	1440	1457	1213

stretching from southern Ghana to southern Republic of Benin, in which the annual rainfall drops below 1200 mm, and in some places below 1000 mm. The main dry season in this area lasts for about four months and a characteristic less dry season from one to three months. These features militate against successful cultivation of several tree crops in this area unless with supplemental irrigation. This area connects in a north-eastern direction with the rainfall regimes of the Guinean savanna with five or more dry months. This zone has long formed a barrier to the spread of tree crop cultivation. The monthly rainfall records at four typical locations within the tree crop belt are given in Table 1.

1.3 TEMPERATURE

The average daily temperatures in this area vary from 23 °C to 30 °C, the lower temperatures occurring from June to November, and coinciding with periods of high humidity. Table 2 indicates the average daily temperatures month by month.

1.4 HUMIDITY

During most of the rainy months relative humidity is high: in the nights and early morning hours relative humidity may be more than 90 per cent. The West African tree crops are adapted to these high relative humidities. During the dry season, especially from December to February, the northern harmattan winds often cause a sharp drop in relative humidity to 40 per cent in the day time, rising again to 60 per cent or more in the early hours of the morning.

Table 2 Average daily temperatures (° C) and their range throughout the year for the West African tree crop belt (van Eijnatten, 1969)

Month	Temperatures		
	Minimum	Average	Maximum
January	20 to 25	26 to 28	29 to 34
February	21 to 24	26 to 29	30 to 34
March	21 to 24	26 to 28	29 to 32
April	21 to 24	26 to 28	29 to 32
May	21 to 24	25 to 28	29 to 32
June	21 to 24	25 to 28	29 to 32
July	21 to 24	23 to 26	26 to 29
August	21 to 24	23 to 26	26 to 29
September	22 to 24	24 to 26	26 to 30
October	22 to 24	24 to 26	26 to 30
November	21 to 24	24 to 27	27 to 32
December	21 to 24	24 to 27	27 to 32

1.5 IRRADIATION AND PHOTOPERIOD

Day length shows a small amplitude in the tree crop belt of West Africa, because the area lies within 8 degrees of latitude from the equator. The difference between the longest and the shortest day lengths is less than one hour. Kola and cacao are, however, cultivated in Jamaica (18 °N) and at Bahia, Brazil (12 °S) and have not exhibited any deleterious effects of differences in day lengths reaching ranges of 10 to 14 hours.

More important are seasonal changes in sunshine and intensity of irradiation. Intensity of irradiation varies considerably with the course of the seasons in West Africa. Mean daily duration of sunshine may be as low as one hour during the months of July and August, the middle of the rainy season. During the dry season, the duration and intensity of irradiation are considerably reduced in many parts by harmattan dust haze. The dust haze originates from the Sahara desert region to the north of West Africa and is intensified by the smoke and ashes from the dry season bush fires in the West African savanna during that time of the year. Daily sunshine duration is longest is the months of February, March and April, when the early rains have cleared the atmosphere, and again in October and November when the rainy season cloud cover diminishes and the harmattan dust has not reached the coastal area. The intensity of irradiation (insolation) is highest throughout the period from February to November and other periods when the cloud cover allows the sun's rays thorugh.

1.6 SOILS

The main part of the tree crop belt is underlain by the pre-Cambrian Basement Complex and the soils are usually derived *in situ* from crystalline rocks or transported over small distances. Only the southern part of the Nigerian tree crop producing area and a narrow coastal strip in Ivory Coast are underlain by tertiary sand deposits (D'Hoore, 1964; Harrison Church, 1960).
Three main soil types can be recognized in the tree crop belt of West Africa:

1. Ferrallitic soils represent final stages of weathering and leaching, and do not contain reserves of weatherable minerals in the form of original rock debris. The silt content is low. The base exchange capacity and its level of saturation are low. The low fertility levels of ferrallitic soils vary with their clay fraction; they may easily be exhausted by cropping and rapidly lose any reserves accumulated by the vegetation in the top layers of the soil.
2. Ferrisols are less weathered, shallower and younger than ferrallitic soils. They have a higher silt content, a higher base exchange capacity, a better structure and higher biological activity. These soils are therefore more fertile than ferrallitic soils.
3. Ferruginous tropical soils develop in areas with a distinct dry and wet season from crystalline rocks of the Basement Complex. The effect of the parent material is marked, as these soils often have appreciable reserves of

weatherable minerals. The soil profiles are seldom thicker than 250 cm and are separated by a thin layer (100 cm or less) of rotten rock from the parent rock. The exchange capacity is low, but higher than that of ferrallitic soils and ferrisols.

In Sierra Leone and south-western Liberia yellowish brown ferrallitic soils predominate. In north-eastern Liberia and south-western Ivory Coast there are yellow and red ferrallitic soils, derived from various parent minerals. In south-central Ivory Coast and south-central Ghana the soils are classified as ferrisols or forest ochrosols, partly interspersed with ferruginous tropical soils. In south-eastern Ivory Coast and south-western Ghana red and yellowish brown ferrallitic soils or oxysols predominate.

The Nigerian tree crop producing area has two distinct zones. The southernmost strip along the coast, up to 60 km in width, has red ferrallitic soils developed on loose sandy sediments. The northern part of the area consists of ferruginous tropical soils, developed from crystalline acid rocks.

The ferrallitic soils of low fertility predominate in the tree crop belt. The production areas of Sierra Leone and Liberia occur on these soils. The areas of high production in Ivory Coast and Ghana are situated on the more fertile ferrisols. The important tree crop producing areas of Nigeria are situated partly on ferrallitic soils and partly on ferruginous tropical soils. The relative importance of the different types of soils in West Africa is shown in Table 3.

1.7 DISTRIBUTION OF THE TREE CROPS

The concentration of tree crops varies greatly within the tree crop zone of West Africa. The main factors which decide the distribution of the various crops within the zone are environmental (Figures 1 and 2). The palms – oil palm and coconuts – have the bulk of their groves along the coast; further inland, the palms give way to rubber in the very high rainfall areas; in the relatively drier areas cacao, kola, coffee and citrus are grown. The areas of the various tree crops overlap to a considerable extent. Also, mixed stands of various tree crops are a common feature, e.g. palms mixed with cacao, kola.

Table 3. Prevalence of various soil types (D'Hoore, 1964) within the West African tree crop belt expressed in percentages

Description	Percentage surface area
Ferruginous tropical soils	8.7
Ferrisols	20.0
Ferrallitic soils	59.9
Other soil types	11.4

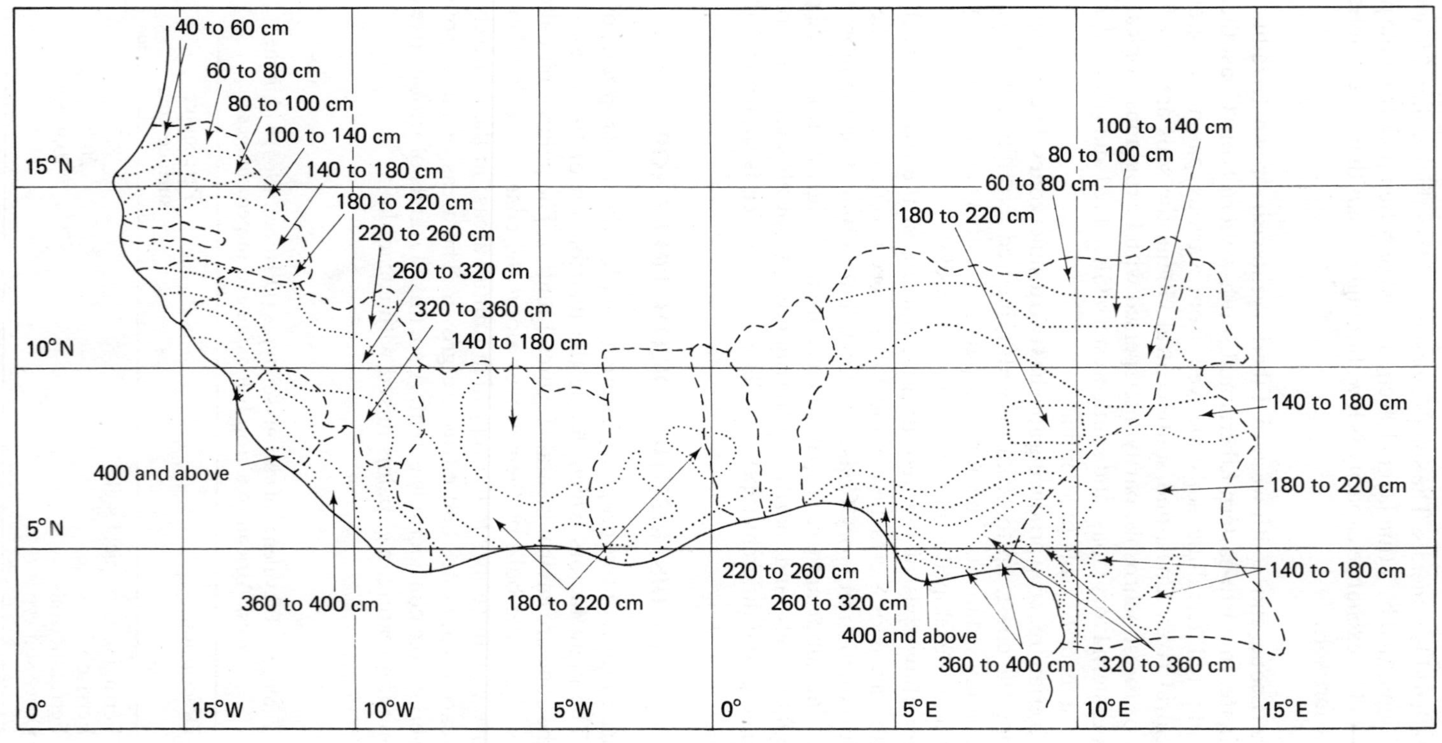

Figure 1. West Africa: annual rainfall

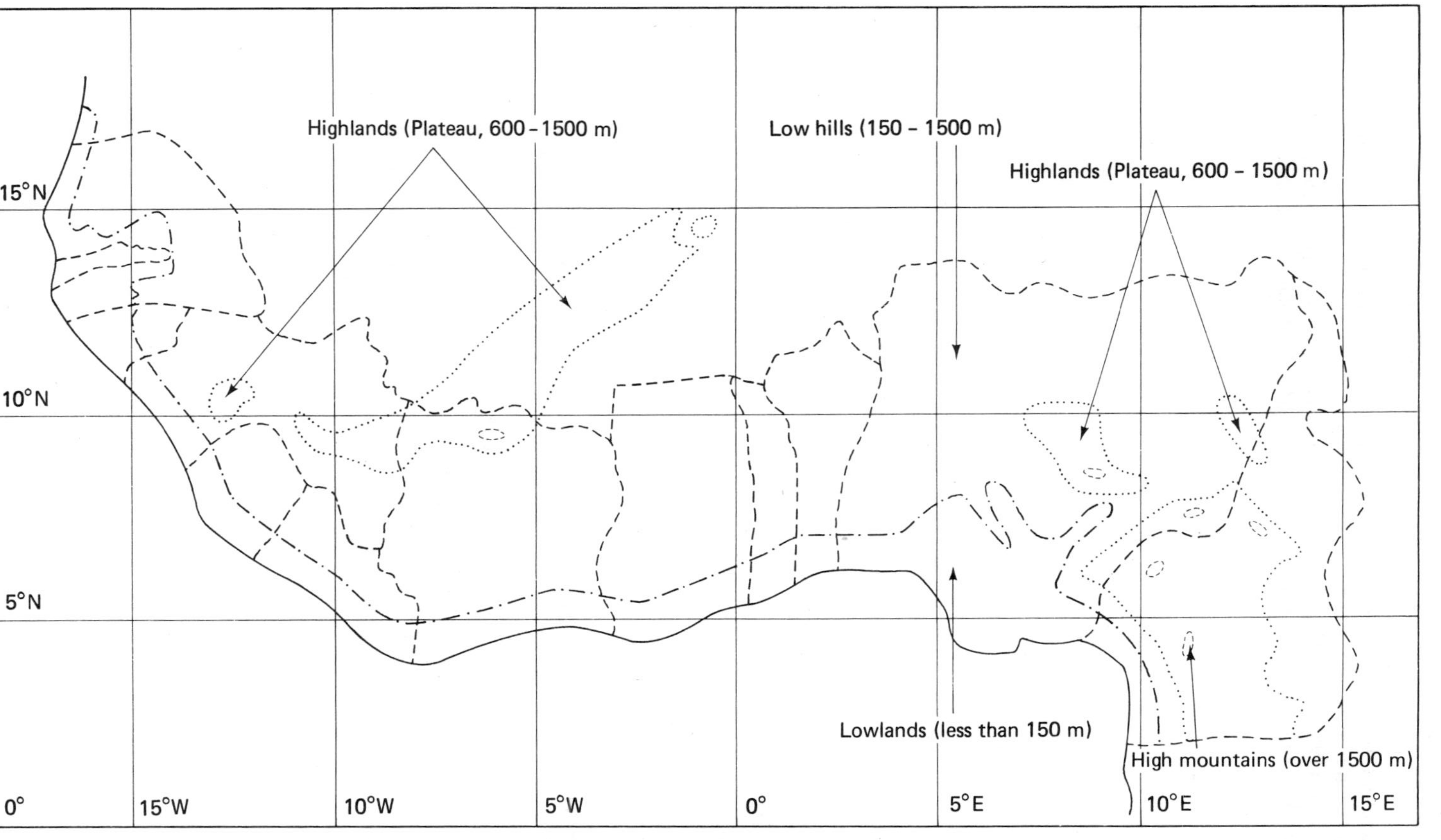

Figure 2. West Africa showing altitude and distribution of major tree crops. Lowlands – mainly coconut, rubber and palms. Low hills – main tree crop area of West Africa: cacao, robusta coffee, cashew, rubber, kola, oil palms, citrus, date palm and 'magic plant'. Highlands and high mountain areas – arabica coffee and tea.

ADDITIONAL READING

van Eijnatten, C. L. M. (1969). *Kola: Its Botany and Cultivation*. Amsterdam.
Harrison, Church (1960). *West Africa*. London.
D'Hoore, J. (1964). Soil Map of Africa. *Comm. Techn. Coop. Africa Publ.*, **93**.
Wills, J. B. (1962). *Agriculture and Land Use in Ghana*. Macmillan, London.

CHAPTER 2

The Nursery

2.1 WHY A NURSERY?

The nursery is the place where the young crop plants are raised under intensive management for later transplanting to the field. Although many of the important tree crops can be sown directly into the field, experience has shown that raising seedlings in the nursery has a number of advantages. Some of these advantages are as follows:

1. Economy of seeds – fewer seeds are required for raising seedlings in the nursery than for sowing directly in the field. For instance, cocoa is sown at stake, two seeds are planted per planting hole, the seedlings later being thinned to one. However, when seedlings are raised in the nursery, only one seedling is planted per stand. For crops such as cacao, coffee, kola, citrus, sour sop, sweet sop and guava, the savings in seeds by raising seedlings in the nursery are usually between 40 and 50 per cent.
2. Seedlings receive more intensive care (protection from animals, diseases and pests, regular maintenance practices, watering/irrigation, manuring, application of artificial fertilizers) in the nursery.
3. Raising seedlings in the nursery affords the planter an opportunity for selecting well-grown, vigorous and disease-free seedlings.
4. The nursery provides the young plants a better medium of growth than when seeds were sown in the field.

Raising seedlings in the nursery followed by transplating to the field also has disadvantages. They concern mainly the high cost which nursery practices introduce into the total cost of crop cultivation. Costs of nursery equipment, tools and materials are additional to the normal costs of field planting. Nursery labour is specialized and therefore expensive. The high cost introduced into permanent tree crop agronomy by nursery practices flows over to field costs. It is more expensive to transplant seedlings than to plant seeds at stake.

Despite these disadvantages most tree crops are established in a nursery, particularly when either the varietal or genetic nature of the material can be guaranteed only for nursery grown/selected material, or when special techniques (e.g. budding) demand nursery techniques.

2.2 SELECTION OF A NURSERY SITE

In choosing a site for the establishment of a nursery, a number of important aspects must be taken into consideration. These can be broadly grouped into two: **environmental factors** – natural features of the nursery site may greatly influence the cost of operation and facilitate management of the nursery, this includes easy accessibility of the site; and **procurable factors** – the availability of various inputs such as chemicals, labour supply and irrigation water is of the greatest importance.

2.2.1 Environmental Factors

2.2.1.1 Nearness to planting site

It is of great importance if nursery sites are located on or very close to the planting site. Some of the advantages include the following:

1. Reduction of the high cost of transportation of seedlings (e.g. transpor-

tation of seedlings with ball of earth, seedlings treated with clay slurry, etc.).

2. Less risk of loss of seedlings during transportation, and seedling failure after transplantation.
3. Limiting or reducing the chances of transmitting or redistributing soil-borne pathogens through seedling roots or earth balls over long distances. When, however, particular diseases occur in an area it may of course be advantageous to raise the seedlings outside the affected area in order to initiate new plantings with disease-free material.
4. Clever coordination of transplanting operations with uprooting of seedlings in the nursery.

2.2.1.2 Water supply

Nurseries should be located where a dependable, abundant and inexpensive supply of uncontaminated water is available. A large volume of water is required daily in the nursery for consumption by plants and workers, for preparation of nursery chemicals and for cleaning. Water supply could be from wells, bore holes, natural streams or irrigation channels. The cheaper the water supply the lower the cost of operating the nursery. Among all factors to be considered in choosing a site for a nursery, a reliable water supply should not be compromised. The supply of 10 mm of water per hectare of nursery beds would require 100,000 litres of water. As evapotranspiration in the tropical lowland is often up to 6 or 7 mm per day, 100,000 litres would only suffice for $1\frac{1}{2}$ days. Water requirements for any additional operation should be added to the requirements of the plants.

2.2.1.3 Land gradient

It is an advantage if level pieces of land are selected as nursery sites. This will reduce the costs of establishing and maintaining the nurseries considerably. Additional advantages areas follows:

1. Reduction in risk of soil erosion.
2. Easy movement of workers, nursery tools and machines around the nursery.
3. Easy application of water through irrigation.

If nurseries must be located on sloping land, soil conservation measures are required, such as bench terracing, contour terracing, or strip planting. The type of conservation measure adopted will depend on the gradient of the land and the nature of the soil.

2.2.1.4 Soils

Soils of nurseries should be chemically and physically fertile or potentially fertile. Favourable soil conditions – good drainage, absence of toxicity, concretions, and excessive clay – are indispensable for the success of a nursery. Where nursery plants are raised in pots, polybags, cane baskets, seed boxes or trays, it may not be necessary for soils on the nursery site to be of high fertility quality. But in this case, a source of high quality soil must be as close to a nursery site as possible in order to lower the cost of soil transportation.

2.2.1.5 Drainage

Whether seedlings are raised directly on nursery soils or in containers – seed boxes, polybags, trays or pots – it is an advantage if the site is well drained. Absence of water pools and muddy spots in the nursery facilitates disease and pest control measures and general upkeep of the nursery.

2.2.1.6 Aspect

Nurseries should preferably not be exposed to winds. Generally, nursery beds are constructed across the direction of any prevailing strong winds. However, some nursery experts construct their nursery beds lengthwise in the direction on winds. In West Africa, data are not available on the relative merits of these two methods of nursery bed orientation. Where nursery beds cannot be located in wind-free sites, permanent windbreaks – plantain, banana, citrus, oil palm – should be planted around the nursery. In addition to serving as windbreaks, these trees will produce harvestable fruits for human consumption and materials for shade and wrapping in the nursery. If nurseries are to be sited in the drier areas north of the West African forest belt, windbreaks should be established with drought-resistant trees such as mango, cashew, neemtree, dongo yaro, datepalm.

2.2.2 Procurable Factors

These are factors which can be supplied to the nursery site, but the nearness of their source of supply is of economic and operational advantage in nursery management.

2.2.2.1 Labour supply

Nursery operations are labour intensive. Therefore, it is very important that nurseries are sited in areas where a dependable and regular supply of experienced or adaptable labour can be easily obtained. One important factor in the management of nursery labour in West Africa is the effect of religious taboos. Expensive seedlings have been ruined because of the refusal of nursery labourers to work on certain days of the week or during periods of religious ceremonies. It is also of great advantage if nursery workers are given regular training and periodic orientation seminars.

2.2.2.2 Market

The nearness of nursery sites to potential buyers is of particular importance to commercial nurseries which raise seedlings for sale to planters. The nursery should be sited as close as possible to these planters.

2.2.2.3 Supplies

There are two categories of supplies to the nursery – these are capital equipment or tools, and material (consumable supplies). The sources of these supplies, especially the consumable ones, should be as close as possible to the nursery site. In this regard, closeness connotes nearness of nursery sites to good roads, railway stations, service centres.

2.2.2.4 Services

Services to the nursery are rendered by agricultural or horticultural experts who are located in the Ministries of Agriculture, the Universities, or the Research Institutes. Nurseries should be located in areas where the services of these experts – pathologists, entomologists, plant nutritionists, agronomists and horticulturists – can be obtained.

Other aspects of services are the availability of good roads and railway stations which are necessary for transportation of supplies, seedlings and workers.

2.3 TYPES OF NURSERIES

In West Africa, there are generally three types of nurseries – the peasant nursery, the intermediate nursery and the standard nursery:

2.3.1 The Peasant Nursery

These are nurseries where peasant farmers raise their own seedlings of kola, cacao, coconuts and other tree crops for planting on the compound or on the farm. Peasant nurseries are located invariably along streams or river banks, along swamps, by the side of family bathroom sheds, or any other permanent source of water. Generally the site is underbrushed leaving trees to provide shade, the soil broken up with the hoe, the seeds sown and covered with palm fronds. In areas where incidence of animal damage to seedlings is probable, the nursery site is fenced round with sticks. Two or more openings are usually left in the fence. Traps are set in these openings for animals that may attempt to invade the nursery. Figure 3 shows a typical peasant nursery.

After seeds have been sown, the nursery is inspected regularly for germination, animal damage or theft. However, generally little care of seedlings or nursery management is practised. The nursery sites are temporary as they shift whenever

Figure 3. Photograph of a peasant nursery

the farmer opens up new forest lands for planting. The main advantage of the peasant nursery is the low cost of the seedlings although these latter often develop poorly.

2.3.2 Intermediate or Temporary Nursery

Intermediate nurseries are improved types of peasant nurseries. They are used to raise seedlings very close to the planting site so as to avoid the cost and problems attendant on long distance transportation of seedlings. Generally, there are no permanent installations in the intermediate nurseries. The site which is cleared, fenced with wire netting and provided with temporary shade, could be used for one season or more. Intermediate nurseries could also be used as resting stations for seedlings which have been transported over long distances. Figure 4 shows an intermediate nursery in Ondo State of Nigeria.

2.3.3 Standard Nursery

The use of standard nurseries dates back to the last decade of the nineteenth century when cacao and oil palm cultivation was vigorously pushed in both British and French dependencies in West Africa. Standard nurseries are expensive but they usually produce better seedlings because growth conditions for plants are better controlled (Figure 5). The layout of a nursery is very

Figure 4. Photograph of an intermediate nursery

Figure 5. Photograph of a standard/permanent nursery

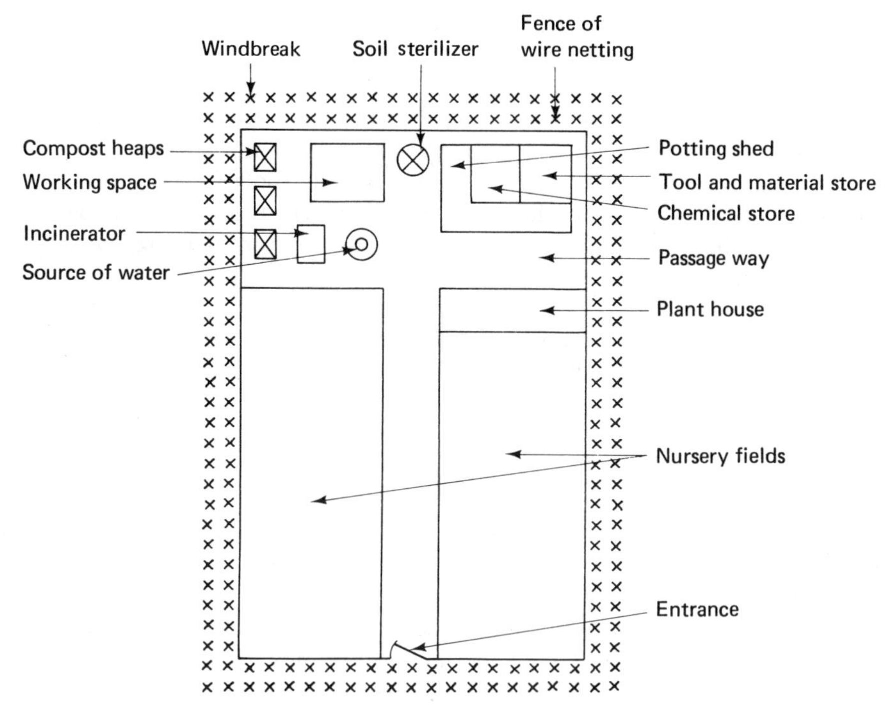

Figure 6. Plan of a standard nursery

important. The layout determines whether a nursery is easy or difficult to operate and will provide good plant material cheaply or at great cost. The standard layout of nurseries recommended in West Africa is illustrated in Figure 6.

Important aspects of the standard nursery are as follows:

1. *Windbreak* With the choice of suitable materials, the windbreak can provide shade materials and sticks for use in the nursery.

2. *Fence* It is advisable that the fence is internal to the windbreak. The fence is usually 1–1.25 m high. Barbed wire should be avoided as material for the fence.

3. *Tools and material shed* This is a walled room provided with a doorway and two windows. All tools, machinery and materials used in the nursery are stored here. Its size will depend on the quantity of tools and materials in use in the nursery.

4. *Chemical store* A walled room, provided with a doorway and two windows one of which opens directly into the potting shed. It lies between the tool shed and the potting shed. Both the tools and chemical stores are roofed with good roofing materials (iron or asbestos sheets), combustible roofing materials should be avoided.

5. *Potting shed* A wire netting or expanded metal fenced area adjoining the chemical store with a verandah space spanning both the chemical and tool stores. The potting shed and the verandah are roofed with transparent polysheets. As the name implies, it is the area where seed boxes, baskets, polybags are filled, seeds sown, seedlings potted, etc. before they are transferred to the plant house.

6. *Compost heaps* Nursery plants grow fast and demand much from the soil, chemically and physically. It is advisable to prepare compost on the nursery site, as nursery refuse provides a reasonable quantity of material. The compost heaps should be adjacent to the incinerator for operational reasons.

7. *Entrance* The main entrance to the nursery should be provided with a lockable iron gate. It should be wide enough to take supply and evacuating vehicles.

8. *Rooting bins* These are propagators, generally specially constructed with concrete for rooting cuttings that cannot be successfully rooted by the polysheet method (see Chapter 3).

9. *Source of water* This could be a well, a tap or a tank that is fed through tanker services. It is advisable that it is located next to the potting shed.

10. *Nursery field* This is where the nursery beds are made or the polybags arranged. The nursery field is generally shaded to the requirement of the seedlings being raised. For this purpose it is advisable to erect permanent supports or durable wood 2.25 m high. Nursery beds are made with pathways 0.50–0.75 m wide. In some cases, the pathways may be planted with tuft (carpet) grass to assist in weed and erosion control. When pathways are grassed, care should be taken to ensure that the grass does not invade the beds.

11. *Incinerator* This is a handy installation for the disposal of nursery trash which cannot be composted. Any ash could be added to the compost heaps.

12. *Plant house* This is roofed with transparent polysheets. This structure houses delicate plants that require intensive care such as hardening.

13. *Permanent supports for shading materials* These are usually wooden supports but other materials, such as concrete posts, may be used.

14. *Soil sterilizer* This is for sterilizing sifted topsoil which is needed as a medium for sowing or planting (see Figure 7).

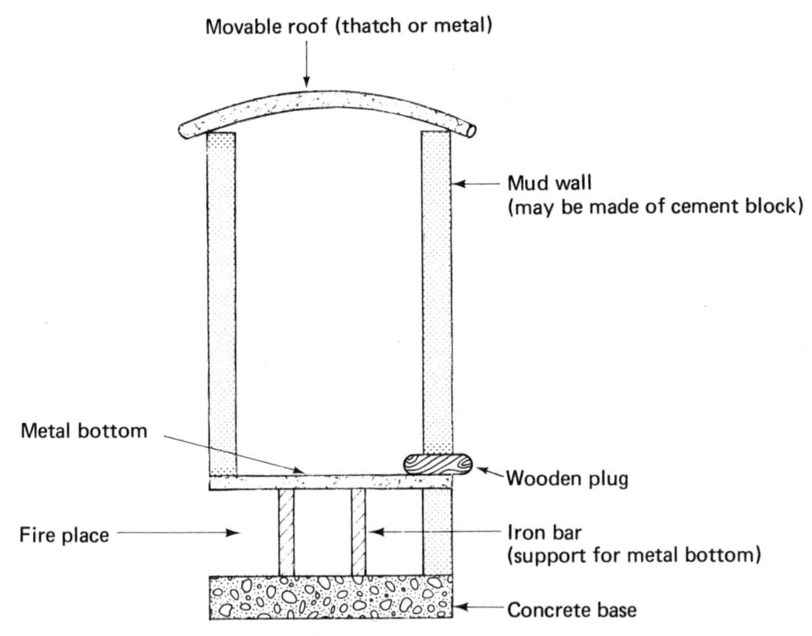

Figure 7. Longitudinal section of a peasant soil sterilizer

Items 1–14 are basic installations in a standard nursery. Additional installations such as fire extinguisher, etc. may be necessary depending on the location of the nursery.

2.4 PREPARATION OF THE NURSERY SITE

The major operations in the preparation of a site as a nursery are as follows:

1. Clearing the site (clear-fell for standard nurseries, but underbrushing and tree thinning for peasant and intermediate nurseries).
2. Removal of the trash by burning followed by removal of debris.
3. Laying out the nursery site according to plan.
4. Planting the windbreaks.
5. Fencing the nursery.
6. Erecting the major installations.
7. Levelling the nursery site (see soil conservation below).
8. Erecting the permanent supports for shading materials.
9. Digging the nursery beds (single digging or double digging). Nowadays polybags are used for raising seedlings and the necessity for digging may not arise.

2.5 SOIL CONSERVATION IN THE NURSERY

2.5.1 Nursery Beds

In addition to providing an appropriate growing medium for seeds and seedings, nursery beds also help to conserve nursery soils from erosion. There are two types of nursery beds.

2.5.1.1 Raised beds

These are usually 1.5 m in width, length as desired. The distinguishing characteristic is that the bed is raised about 10–15 cm above the level of the pathway. Raised bed are used in medium to heavy rainfall areas where bed drainage is needed.

2.5.1.2 Sunken beds

Sunken beds are usually 1.5 m in width, length as desired. The distinguishing feature is that the beds are sunken 10–15 cm below the level of the pathways. Sunken beds are used where it is necessary to conserve moisture – low rainfall areas, areas where moisture supply to the nursery is through irrigation water. Sunken beds are particularly effective as anti-wind erosion measures.

2.5.2 Terracing

Terracing is very important in the humid tropical parts of West Africa on account of the undulating or sloping nature of the land, heavy rainfall, high erodability and consequent loss of mineral nutrients and physical structure of the soils. The type of terracing adopted will depend on the degree of slope of the land.

2.5.2.1 Bench terracing

Bench terracing is most commonly used for erosion control on steep slopes. Terraces are constructed along the contour line of the land (see Figure 8).

Each terrace is supported by a wall or bank of well beaten earth or stone with the terrace sloping gently backwards (uphill) to reduce the risk of crumbling or sliding at a later date. The top of the wall is planted with a slow spreading fibrous rooted grass.

Each terrace is provided with a drain. The drain is located at the bottom of the retaining wall. The drains are constructed in such a way as to ensure that the surface water runs from one drain to the base of the next wall. Drains are connected with other drains at the end of each terrace. This device ensures that during heavy rains, water will not be able to run in a downward direction, carrying silt and topsoil away, but trickles to the back of the terrace and hence to the side drains in which silt catchers are inserted at intervals (see Figure 9).

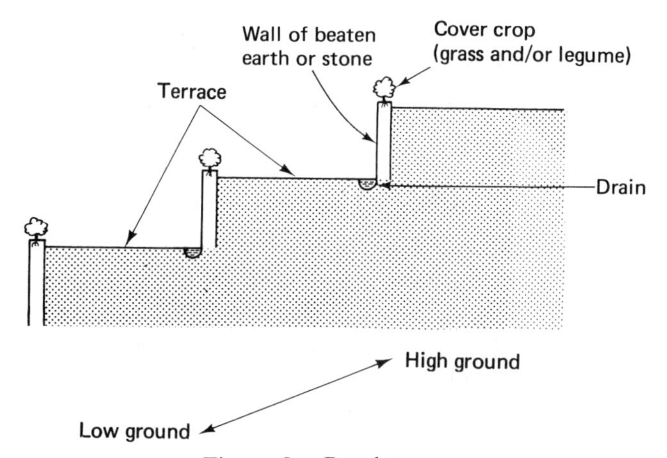

Figure 8. Bench terraces

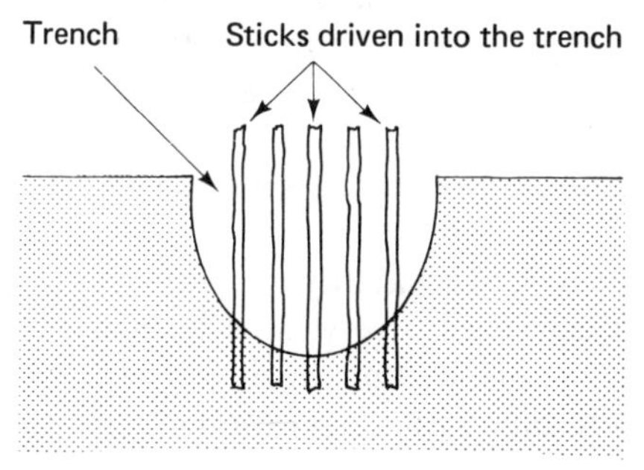

Figure 9. Silt catcher

2.5.2.2 Contour terracing

Contour terracing is similar to bench terracing but the construction is less formal (see Figure 10). The terrace walls are built to conform with the contour lines of the land. Levelling is kept to the barest minimum, and most of this is left to rain water.

Bench and contour terraces are the two most popular soil conservation

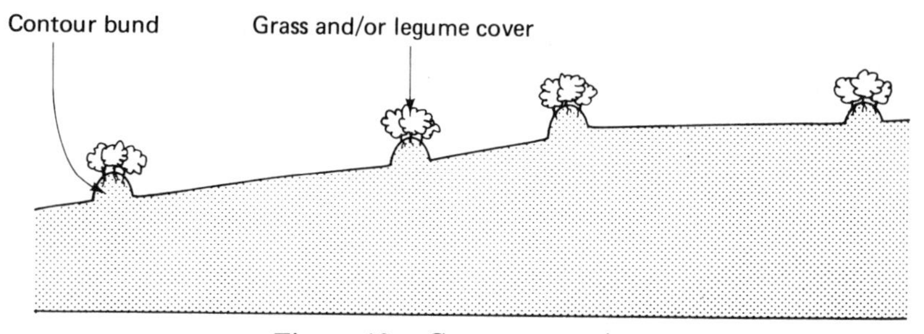

Figure 10. Contour terracing

methods used in nurseries for tree crops in West Africa. They are supplemented with soil traps (see above), cover crops such as *Pueraria phaseoloides*, *Centrosema pubescens*, *Calopogonium mucunoides* on terrace walls or bunds, and effective windbreaks, hedges or shelter belts.

2.6 NURSERY MANAGEMENT

Many valuable tree crop nurseries in West Africa fail on account of poor management. The essential aspects of management in a nursery are as follows:

1. *Cleanliness* This, in addition to making the nursery attractive to workers, helps to control diseases and pests. Cleanliness is achieved through regular weed control, removal of trash and other refuse, use of the incinerator and proper care of nursery materials and tools.

2. *Care of nursery tools and equipment* Nursery tools and equipment must be cleaned after use and properly stored. Metallic tools that are liable to rust must be coated with grease or used engine oil before storage. The tool shed must be properly organized and store items well labelled. The nursery is a unit in the farm enterprise, therefore a separate inventory must be kept for it.

3. *Layout of nursery* The workability of a nursery is greatly influenced by its layout. In other words, the nursery must be laid out to function. Nursery operations are usually in sequence, and the site for each operation in the sequence must be located in accordance with the position of the operation in the sequence. Orderly arrangement of the nursery layout greatly facilitates the operations in the nursery.

4. *Personnel management* Nursery labour is specialized and this makes it very important to ensure that nursery labourers, attendants and supervisors are carefully managed. A number of nursery operations involve soiling of hands, cloths, etc. Nursery workers invariably do not take kindly to performing these operations without protection. Provision of gloves, aprons, overcoats, etc. for nursery workers is indispensable to the success of any nursery.

5. *First aid* The provision of a first aid box in the nursery is a necessity. The nursery labour force must include somebody who has training in first aid especially in treating fresh wounds, snake bites and the like.

2.7 SHADE REQUIREMENT AND MANAGEMENT IN NURSERIES

For most tree crop nurseries, shade is very important. Lack of shade or excess shade where needed will invariably result in poor growth of seedlings. In most parts of West Africa, nursery shade is provided by palm fronds – compound leaves of *Elaeis guineensis* or *Raphia royale*. In some areas, these palms occur naturally and in large numbers. Whether they occur abundantly naturally or not, it is an asset to a nursery to have its planted palm grove from which shade material can be obtained. In some cases, palm trees are used as windbreaks, in which case the palm trees serve a dual purpose – windbreak and supply of shade material. As shown in Figure 5, permanent supports are erected in the nursery for shade purposes. As to cross-stricks over the poles, there are two schools of thought, one holds that there should be permanent cross-sticks while the other recommends annual changes of cross-sticks.

Shade management in the nursery is complicated by a number of factors, the major ones being the frequency of overcast skies and shadow cast either from the surrounding windbreak, the adjoining bush or other structures in the nursery. Generally, shade regimes for crops requiring shade in the nursery are as follows:

1. At planting and up to full development of the photosynthetic apparatus by the seedlings – 50 per cent shade regime, i.e. shade is made to cut off 50 per cent of the incident light. The duration of this period will vary with each crop.
2. After full development of the photosynthetic apparatus, shade for growth is required. This is known as 'growth-shade-period' and it is the longest of the different shade regimes in the nursery. The shade requirement at this stage is usually 60 per cent, i.e. 40 per cent of the incident light is cut off or obstructed by artificial shade.
3. Whether the crop needs shade in the field or not, it is of advantage to harden the seedlings preparatory to transplanting into the field. Adjustment of shade for hardening purposes usually commences about 45 days to transplanting. The sequence of operation in shade reduction for the purpose of hardening is as follows:
   (a) 45 days to transplanting reduce shade to 30 per cent;
   (b) 20 days to transplanting reduce shade to 15 per cent;
   (c) 5 days to transplanting reduce shade to 0 per cent.

In shade management in the nursery, it is absolutely essential to maintain evenness of shade. For sensitive photo-experiments, plastic netting is used. The disadvantages of plastic netting are high expense and early perishability. The main advantages are ease of handling and uniformity of shading or light distribution.

The percentage shade regimes should be adjusted for local weather conditions. The experienced nursery man will easily detect when he has too much light or too dense shade.

2.8 USE OF CLAY SLURRY IN THE NURSERY

Clay slurry is a thick dispersion of fine clay in water. Clay slurry is an important material in the nursery. It is extensively used to protect seedling roots that are transplanted naked from dehydration. Clay particles do not remain for long in suspension in water, but settle down fairly rapidly. When clay slurry is to be used, the whole suspension should be thoroughly and regularly stirred before and during use.

When seedlings are uprooted 'naked root', the naked roots are dipped in clay slurry. Seedlings treated with clay slurry are packed in sacks or palm fronds for transportation to the planting site. On arrival at the site, the bundle is loosened, the seedlings spread out and moistened under shade. Field transplanting should be carried out within the shortest possible time from arrival of the plant materials on the planting site, and in any case within one or two days.

As pointed out earlier, there is still a school of thought which advocates sowing seeds straight (sowing at stake) into the field, thus bypassing the high cost of running a nursery. This attitude shows that more research is needed in the two approaches of establishing tree crop seedlings in the field, namely, raising seedlings in the nursery followed by transplanting and direct seeding in the field. Research is also needed in the area of developing improved field techniques that would make seedlings which develop from direct seeding in the field, grow as fast as seedlings raised in the nursery. When such a technology is developed, its advantages will have to be weighed against the advantages of raising seedlings in the nursery. So far, the use of a nursery is standard practice in crop culture in West Africa.

ADDITIONAL READING

Quarcoo, T. (1973). *A Handbook on Kola.* Cocoa Research Institute of Nigeria, Ibadan, Nigeria.

Tindall, H. D. (1965). *Fruits and Vegetables in West Africa.* FAO Publ. Rome.

CHAPTER 3

Methods of Propagation

The propagation of tree crops depends on the characteristics of the crop concerned. Some are propagated sexually through seed resulting from fertilized egg cells; others may be propagated vegetatively. These methods are not mutually exclusive and some tree crops may be propagated either sexually or vegetatively.

3.1 SEXUAL PROPAGATION

This is propagation through pollination, fertilization (fusion of male and female gametes) followed by seed and/or fruit development, fruit ripening and dispersal. In sexual propagation, the zygotic embryo is the end product of a sexual cycle which involves flowering, formation of gametes, maturation of gametes, coming together of male and female gametes (pollination) fusion of male and female gametes (fertilization) and seed development. This process is illustrated in Figure 11. Zygotic embryos are the normal process of seed production in most West African tree crops.

Non-zygotic embryos may arise through a secondary stimulus resulting from

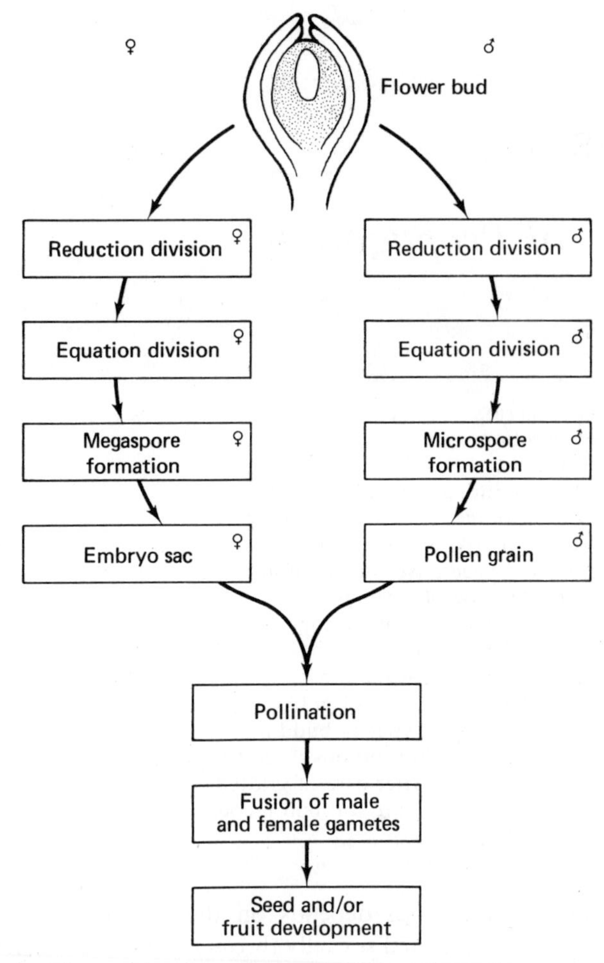

Figure 11. Sexual reproductive cycle in plants

the act of fertilization. When cells of the nucellus develop into an embryo in this manner, **nucellar embryos** are produced. Nucellar embryos are diploid, but all their chromosomes are maternal in origin. This is of frequent occurrence in citrus. When any of the antipodal cells (the synergids) develop into embryos in this manner, **haploid embryos** are produced. These occur occasionally in cacao. These haploid embryos derive their haploid chromosome set from the female parent and they are useful in the production of isogenic lines. Both processes are forms of apomixis and may lead to more than one embryo per seed, i.e. polyembryony.

3.1.1 Seed Selection

There are two aspects to seed selection in tree crops. The first is related to selection of seed for genetic superiority through selection methods applied to the parent stock. In inbreeding plants, this selection method is very efficient, but in outbreeders, and most tropical tree crops are outbreeding, this method of selection is only valuable in the selection of female parents except where hand pollination or controlled pollination is carried out. The essential criteria for seed selection through the parental stocks are that parents should be selected on the following bases:

1. Good growth and conformation.
2. High yielding ability.
3. Freedom from diseases and pests, especially virus and bacterial infections.
4. Good quality (flavour, seed size) of produce.

It must always be remembered that reproduction through seeds does not guarantee the transfer of high performance of parents to offsprings especially when pollination is uncontrolled.

The second aspect is the selection of seed for good agronomic character – this is selection of the best seeds from those available. The seeds to be planted are selected for size (the bigger the seed the better), purity and trueness to type, freedom from pests and diseases and viability.

3.1.2 Seed Viability

Seeds vary greatly in their ability to germinate immediately when they are physiologically mature. Seeds of the citrus and cacao germinate easily when freshly harvested but lose their viability very rapidly after harvesting. They do not stand drying or prolonged storage.

On the other hand, seeds of kola, coffee, rubber, pawpaw and the palms undergo varying periods of post-ripening dormancy. The golden rule is to test seeds for their ability to germinate promptly before planting.

3.1.2.1 Germination and viability tests

Viability in seeds is the possession of the ability to germinate and to produce a normal seedling. Germination may be postponed by dormancy of the seed. Seed dormancy is non-germination of viable seeds, the causes of non-germination being due to either mechanical barriers (hard testa or shell) or internal physiological barriers. The nature of any dormancy should be clear: hard seeds or dormant seeds (*sensu stricto*):

1. **Hard seeds** are seeds which are viable but due to their possession of hard covers they cannot easily absorb moisture for early germination. Examples include seeds of oil palm and rubber.
2. **Dormant seeds** are seeds which are able to absorb moisture very readily but which fail to germinate because of restrictive influences within the seed that block physiological reactions in the embryos and thereby prevent the

initiation of the processes of germination, e.g. physiological immaturity, presence of toxic chemicals and growth inhibitors.

There are many methods for exciting dormant embryos to initiate germination. The most common of these methods are as follows:

1. Mechanical scarification – this can be achieved by passing the seeds through rough metallic surfaces which cause bruises on the seed coat, or in the case of very small seeds, by use of sand paper.
2. Soaking of the dormant seeds in water for brief periods (6 to 24 hours).
3. Acid scarification; the most popular acid in use is concentrated sulphuric acid. Care must be taken to avoid acid burns on the embryos and the cotyledons.
4. Germination under raised temperatures.
5. Moist chilling – not common with tropical crops.
6. Moist heating or hot water shock – this is achieved by subjecting moistened seeds to varying periods of high temperature, e.g. oil palm seeds.

Seed testing for viability is important to avoid planting dead seeds, and to determine proper seed rates. There are three principal methods for testing seed viability.

1. Germination tests
2. Flotation
3. Colouring tests.

Germination tests form the most reliable method. The procedure is as follows:

(a) Obtain a representative and uniform sample of the entire seed lot.
(b) Count 25, 50, 100, or more seeds from the representative sample.
(c) Sow the seeds in seed boxes, baskets or polythene bags filled with sterilized soil. Small seeds could be arranged between folds of moist blotting paper or cloth and be placed between two plates, petri dishes.
(d) Keep the sown seeds at room temperature.
(e) Moisten as often as possible just to keep the seeds moist and *not* wet.

Viable seeds will sprout more or less at the same time. Weak seeds usually sprout irregularly and they do not usually produce normal seedlings. For tree crops, it is normal practice to take records of germination for a period of 8 to 10 days before the germination percentage is calculated.

Flotation of the seed is often reliable, but it is not as sensitive as the germination method. A good sample of the seeds is poured into water contained in a bucket or pot. Viable seeds will sink while non-viable seeds float. This method is very rapid and is recommended when seeds are to be purchased from the open market.

The colouring or Tetrazolium test is a biochemical test through which viable seeds can be identified. The test is carried out by soaking seeds in 2, 3, 5-triphenal tetrazolium chloride (TTC). Viable seeds absorb this chemical into their living tissue where it is changed into an insolubled red compound known as Formassan: non-living tissues cannot effect this change. The TTC test is valuable in distinguishing between dead and live seeds. The reaction takes place equally well in dormant as in non-dormant seeds.

3.1.2.2 Storage of seed for planting

Seeds are living organisms. The objective of storing seeds for planting is to retain viability of the seeds for the purpose of raising a succeeding generation of the crop concerned.

Although it is not often realized that seed storage commences immediately the seeds are mature, some seeds can be successfully germinated immediately upon harvesting. In such a case, the seed can be sown straight after ripening without any deleterious effects. In other cases, physical seed maturity precedes the physiological maturity of the embryo. In such a case, the seed (embryo) has to undergo a period of post-ripening maturation (dormancy period). It is not desirable to break this type of dormancy through artificial means and seeds need to be stored at least for the duration of the dormancy period. There are a few cases where the maturity time of the seed and the embryo coincides, but as the seedman terms it, the embryo 'went asleep' immediately after maturation. The embryo takes a rest. If seeds containing such embryos are sown immediately after maturation, they do not germinate promptly. For good vigour of growth to be achieved, the rest period should be allowed. The seeds should be stored. If the necessity, however, arises, the embryos could be incited to resume physiological activity of growth through artificial means. In addition, there are a number of agricultural reasons for seed storage. Among these are the following:

1. Seeds may ripen outside the planting (sowing) season of the crop.
2. Seeds may need to be transported over long distances.
3. A particular batch of seeds may need to be preserved for repeated plantings for the maintenance of the gene pool of the specific seed stock over a number of months or years.

Seeds of tropical tree crops are of four types storagewise:

1. Wet seeds – citrus seeds, cacao beans, avocado pear.
2. Semi-wet seeds – kolanuts, breadnuts, mangoseeds.
3. Dry/wet drupes – coconuts, palm kernel.
4. Dry seeds – sour sop, sweet sop, coffee.

These different categories of seeds require special storage techniques.

3.2 ASEXUAL OR VEGETATIVE PROPAGATION

Asexual propagation is reproduction from vegetative parts of plants. This is possible because the vegetative parts (organs) of many plants have the capacity for growth when appropriately treated, e.g. stem cuttings, root cuttings, leaf or single bud cuttings. Also single cells can be used to re-establish a plant through appropriate tissue or cell culture techniques.

Asexual propagation results in the production of genetically uniform material derived from a single individual and propagated exclusively by vegetative means. Such material is known as a 'clone'. Clones may originate from seeding material or from a budsport.

The main advantages and uses of vegetative propagation include the following:

1. Preservation of genetic characteristics, with the exception of variability which arises through mutations.
2. Production of clonal materials which are generally early maturing.
3. Provision of a technique for raising peculiar seedlings, e.g. rootless seedlings can be raised to mature plants by grafting them on to compatible root stocks.
4. Control of mature plant size, as clonal plants are more uniform than seedling plants, while actual size can be influenced by the use of vigorous or dwarfing root stocks.
5. Where grafting/budding is involved, the stock/scion may have favourable effects on one another or their interaction improve the product, e.g. flavour or size of fruit in some species of citrus.
6. Vegetative propagation provides a means for repairing damage to trees or changing of varieties in a plantation.
7. Vegetative propagation techniques are useful in disease studies such as 'virus indexing'.

The main disadvantages of vegetative propagation are the following:

1. The method is more costly than propagation through seeds because special techniques are to be applied, often time consuming, and requiring various constructions and equipment (propagation bins, shade houses, irrigation equipment).
2. If not carefully controlled, virus and bacterial diseases can be spread through vegetative propagation.
3. Often, problems of graft or bud – incompatibility – make budding or grafting impossible or at least extremely difficult.
4. When it involves rooting, some plants turn out to be difficult to root.

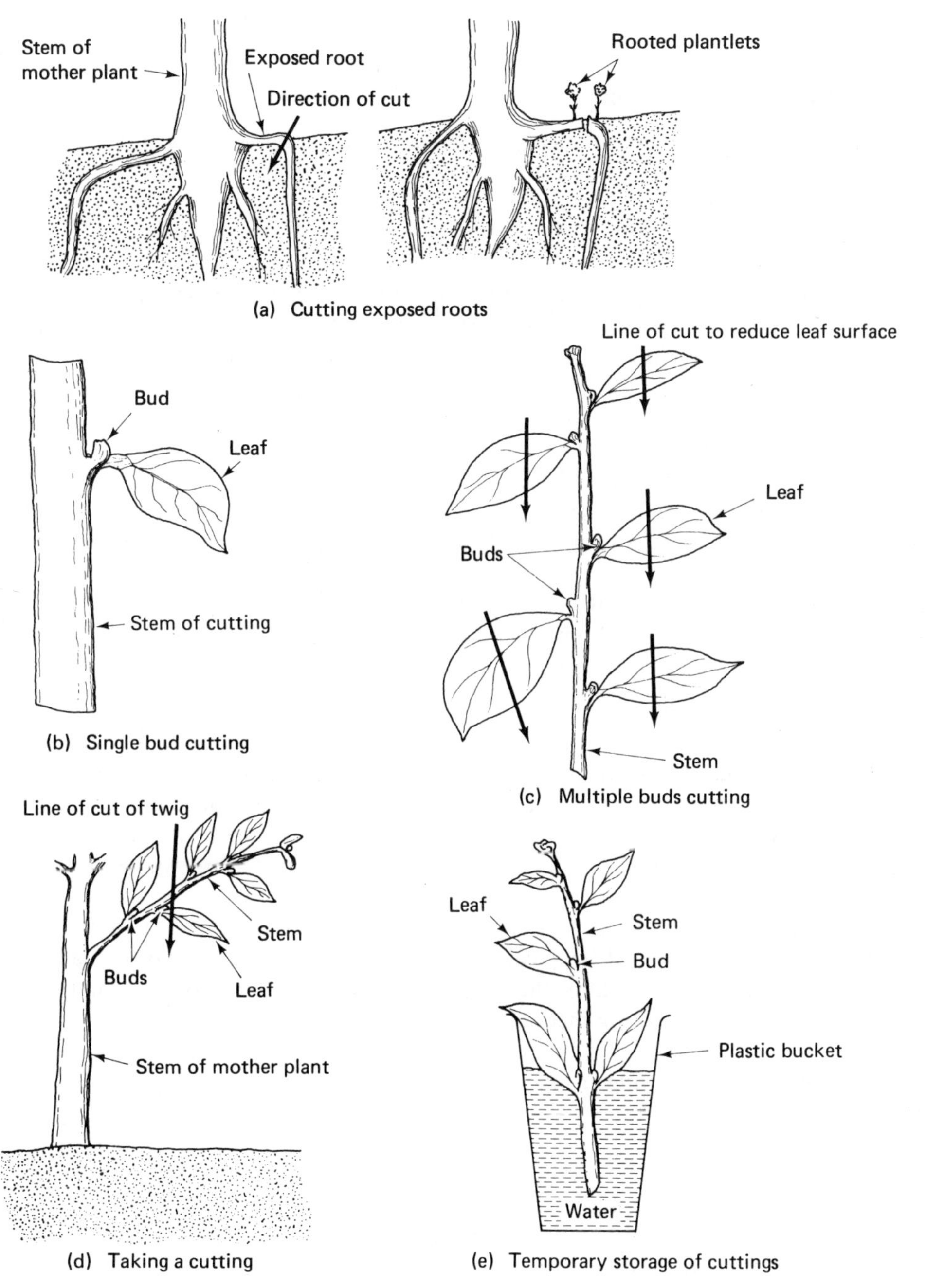

Figure 12. Various methods of rooting plant material

3.2.1 Methods of Vegetative Propagation

3.2.1.1 The use of cuttings

There are various types of cuttings. The most important types as applicable to tree crops in West Africa are briefly discussed below:

3.2.1.1a Root cuttings

The most popular method of root cuttings in practice with tropical tree crops is the '*in situ* root cutting method'. This method is used commonly with the breadfruit (*Artocarpus altilis*) and the jakfruit (*heterophyllus*). It is also used with sweet orange by the peasant farmers of West Africa. Cutting of the root could be replaced by wounding.

In practice, the root which is to form young plants is exposed and cut across or wounded and left *in situ* (Figure 12a). This is done during the rainy season. New growth may be stimulated by the application of growth substances. With the breadfruit, shoot primordia and root primordia appear about three to four weeks after cutting. Both proximal and distal parts of the roots may develop plantlets. When the plantlets have developed sufficiently, they are cut, potted and nursed until they are ready for transplanting into the field.

3.2.1.1b Stem cuttings

1. *Leaf bud cutting* (*single bud cutting*) This type of cutting consists of a leaf blade, petiole and a short stem piece with an axillary bud. This method is used in species where viable buds are scarce or in species where the commercial product consists of this source material as in tea (Figure 12b).

2. *Multiple Bud cutting or stem cutting* These are used with most of the dicotyledonous species of tropical tree crops. The practice involves taking semi-mature (6 to 12 months old) twigs carrying viable buds and rooting them in an appropriate rooting medium (Figure 12c, d and e).

3. *Procedures for rooting stem cuttings* Materials required for rooting or stem cuttings are as follows:

(a) A rooting bin (concrete, boxes or polythene sheet) (see Figure 13) and polypots or cane baskets.
(b) Rooting medium (sterilized soil, sawdust mixed at desired ratio).
(c) Rooting hormones where needed. The most popular rooting hormones currently in use in West Africa are indole acetic acid (IAA), indole butyric acid (IBA) and the gibberellins.
(d) Budding or grafting knives.
(e) Plastic buckets, polythene bags.

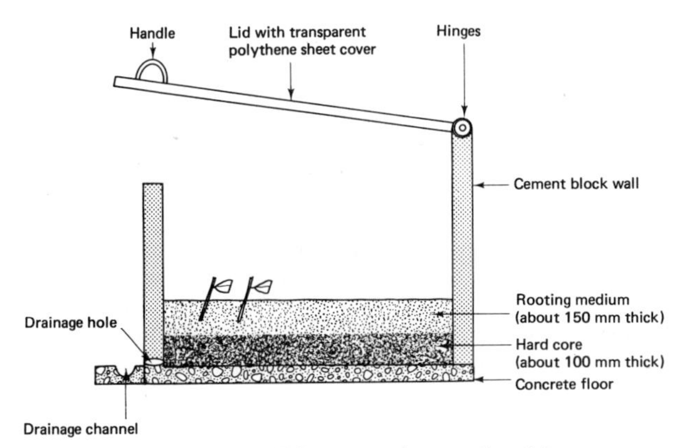

Figure 13. Diagram of a rooting bin

(f) Working table or platform.
(g) Water.
(h) Budding tapes or raffia.

The process of rooting evolves as follows. The rooting medium is prepared by filling either the rooting bins, the polypots, or cane baskets with the appropriate sterilized rooting medium. This is well watered. It is advisable to carry out this operation the day preceding setting of cuttings. Very early in the morning of the day on which cuttings are to be set, twigs of the crop to be rooted are taken and kept in water in a plastic bag or plastic bucket (see Figure 12e). They are conveyed to the rooting site. At the rooting site, the plastic buckets containing the twigs dipped in water are placed on the working platform. Other materials required for rooting are assembled on the working table/platform. After all these preparations for rooting have been satisfactorily completed, the actual rooting operation follows immediately. The operation consists of the following steps:

(a) Cutting a fresh surface of the 'twig' to be rooted, under water.
(b) Gently lifting the 'twig' and inserting it firmly in the medium, dipping the cut tip of the twig in the appropriate rooting hormone where necessary before insertion. It will be observed that as the cut twig is gently being taken out of the water, a drop of water will hang from the very tip of the cut surface. This drop of water is a continuation of the film of water which protects the cut surface of the twig from air penetration. This drop of water must be preserved until the twig is inserted. Where the twig is treated with rooting hormone, the drop of water is replaced with a drop of rooting hormone. When powdered rooting hormones are used, the drop of water is lost, but the film of water on the cut surface binds the powdered hormone tightly. Nevertheless, the time interval between dipping the twig in the hormone and insertion in the rooting medium should be kept at the barest minimum.

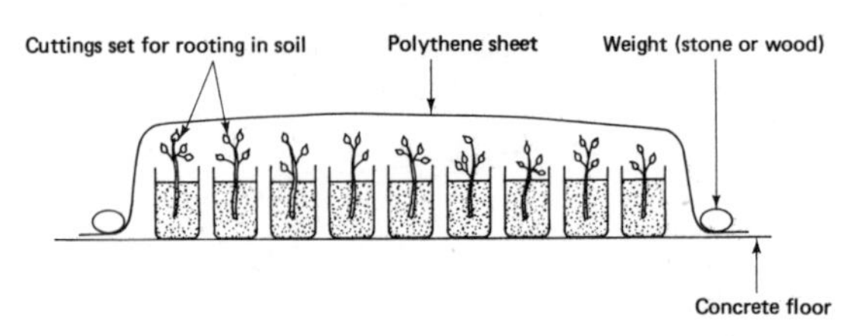

Figure 14. Polythene sheet method of protecting rooted cuttings

(c) Consolidation of the soil/sawdust medium around the inserted twig to ensure that air pockets do not develop around the cut surface of the inserted twig.
(d) Watering – this may be slight or heavy but it should not flood the rooting medium
(e) Covering of the cuttings:
  (i) In the case of the rooting bins the covering lids of the bins are to be replaced.
  (ii) An alternative is to cover the set-material with a polythene sheet. The most commonly used polythene sheets are the colourless ones (see Figure 14).
(f) The 'set' is watered regularly until the cuttings show signs of successful rooting. After successful rooting, the rooted material, known as ramets in kola, is potted and hardened.

4. *Potting and hardening* When roots have developed to about 30–50 mm on the rooted cuttings, these are ready for potting, unless cuttings have been rooted directly in polythene bags or cane baskets filled with fertile topsoil. When set in rooting bins, the cuttings are gently lifted and examined for good root formation. Once it is ascertained that the roots are well formed, the roots (very fragile at this stage) are inserted in a hole made in the centre of the soil in the polythene bag or cane basket, to which the cutting is to be transferred (see Figure 15). Poorly developed cuttings are discarded at this stage. The potting medium (the soil) is firmly consolidated around the cutting to provide support. The plants are thoroughly watered and returned to the rooting bin or hardening chamber, but where necessary separate hardening chambers could be constructed. The normal procedure is to place the potted cuttings in the hardening chamber after thorough watering; the chamber is closed for the length of the period required by the crop (see individual crops) without watering. After the period of complete closure, the covers are progressively lifted for about 2–3 cm at a time with only a little watering until the lids are completely lifted at about three to four weeks, which is the end of the hardening period. Watering during the hardening period should be restricted.

If the cuttings were set directly in polypots or cane baskets, repotting is unnecessary. The process of hardening consists of stoppage of watering and progressive and gentle lifting of the sheet for a period of two to three weeks

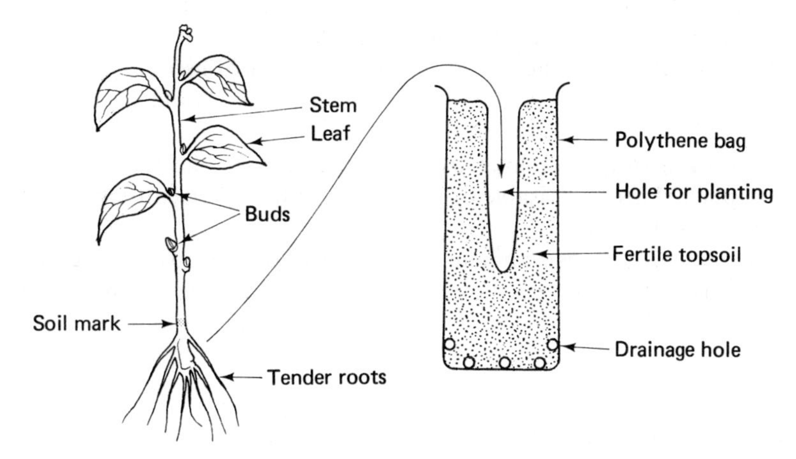

(a) Successfully rooted cutting and polythene bag filled with fertile topsoil

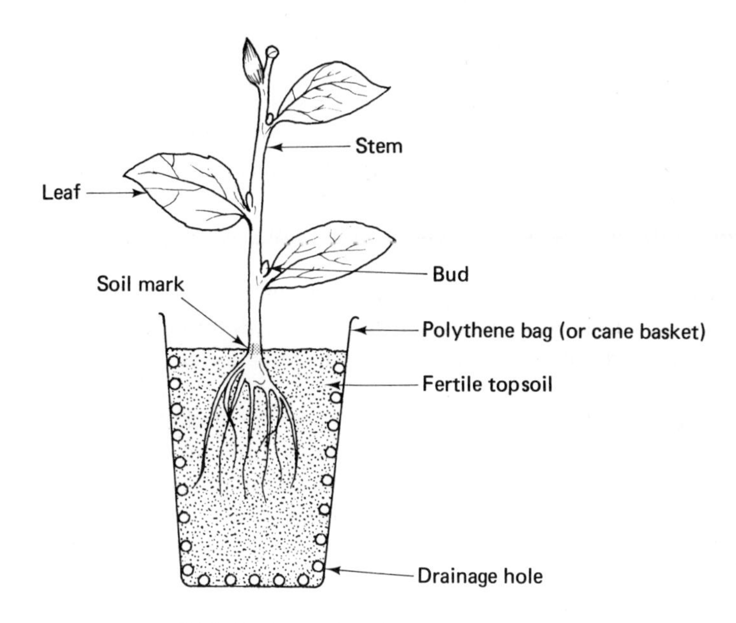

(b) Potted rooted cutting

Figure 15. Transplanting of rooted cuttings from the rooting medium into pots

when the sheets are completely exposed. The hardening process is much simpler in this case, although the principle of the gradual and gentle introduction of the newly rooted cutting into ambient conditions is the same. After hardening, the cuttings should be transferred to shaded nursery field beds. Daily watering, shade adjustment, disease and pest control measures should be carried out regularly.

If cuttings are stored for long periods in the nursery field, some fertilizer may be applied. This is advisable one to two months before cuttings are finally planted in the field.

In some plants, cuttings commence flowering prior to transplanting into the field. Such precocious flowers should be removed, since the cuttings have not yet developed sufficient roots to enable them to support flowering and fruit bearing.

3.2.1.2 Grafting and budding

When plant parts are joined together they are said to be grafted. In strict terms, grafting is used in vegetative propagation when the graft-scion carries more than one bud. When the graft-scion carries only one bud, the term budding is used. Apart from artificial grafting and budding, natural grafting of branches and roots of trees respectively does occur in nature.

An essential requirement for grafting is the availability of root stocks; these are plants which serve as roots for the grafted or budded material. The root stocks could be seedlings or vegetatively propagated clones. The advantages of seedling root stocks are that they are cheap to produce, generally virus free, and deeper rooting thus providing better anchorage. They are, however, genetically variable, and lead to variability in growth and performance of grafted plants.

Clonal root stocks are in popular use in the temperate countries, for example, for apples. They are genetically uniform, with specific growth characteristics, but if not carefully indexed, they may become a channel for the spread of virus infections.

Sometimes inter-stocks are required. There are stocks which are inserted between root stocks and scions for special purposes. Inter-stocks are used to circumvent incompatibility or to take advantage of a specific characteristic of the inter-stock (disease resistance, stimulation of profuse flowering or favourable effects on the quality of produce).

3.2.1.2a Types of grafts

There are many types of grafts. The most common ones are listed (see also Figure 16).

1. Whip, tongue or saddle graft
2. Splice graft

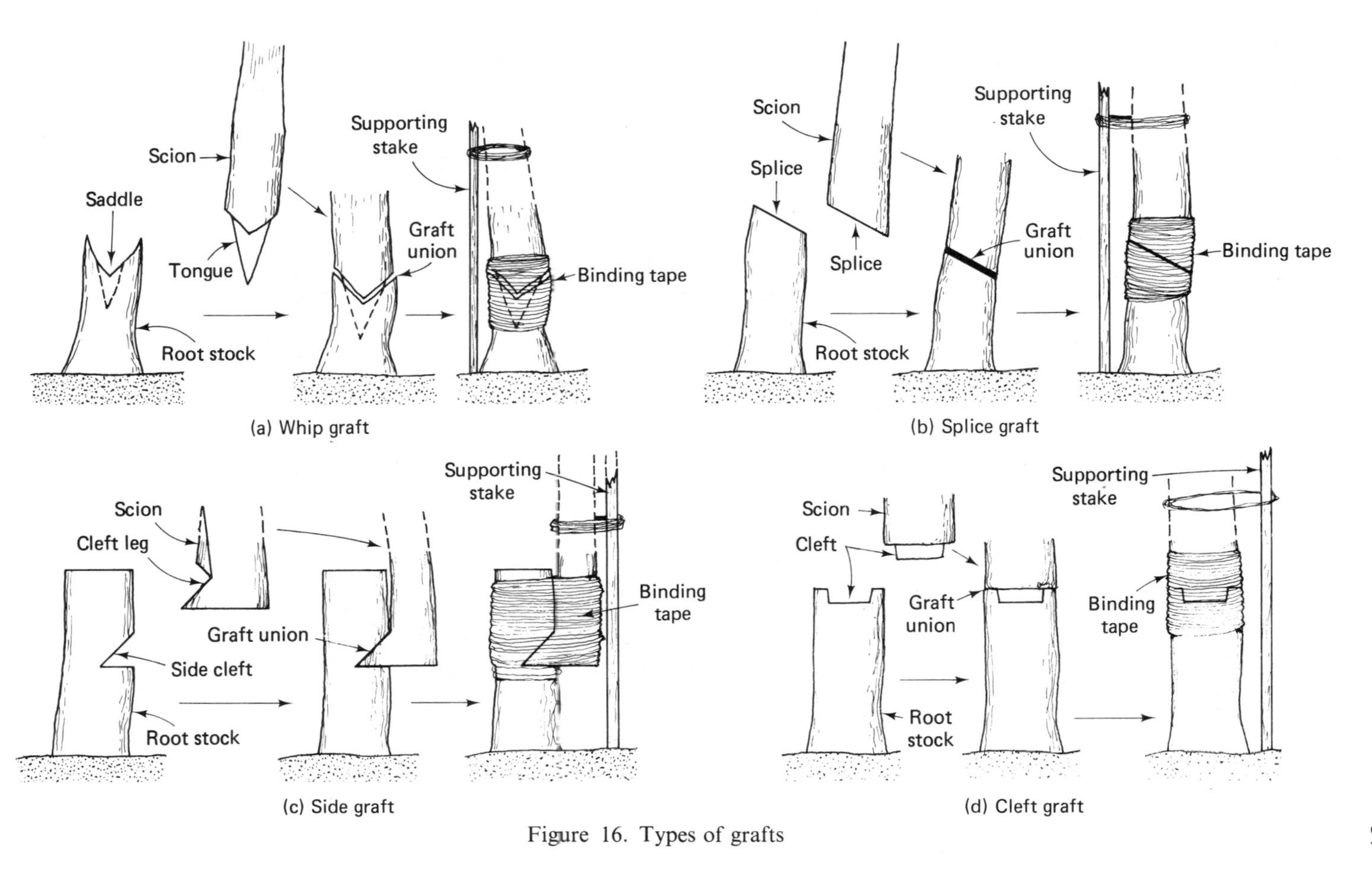

Figure 16. Types of grafts

3. Side graft
   (a) Stub graft
   (b) Side tongue graft
   (c) Side veneer (splice side graft)
4. Cleft graft; this is popularly used in top working
5. Saw-kerf (Hotch graft)
6. Bark-graft
7. Approach graft
8. Inarching (repair graft)
9. Bridge graft (repair graft)
10. Brace graft (bracing – repair graft).

3.2.1.2b Types of buddings

Budding is a form of grafting in which only one bud is involved instead of a shoot with several buds. Several types of budding are listed below (see also Figure 17).

1. **Inverted 'T' budding**, commonly used for citrus.
2. **'T' or shield budding**, similar to inverted 'T' budding with the only proviso that the 'T' is upright.
3. **Patch budding**, mostly used for rubber, mangoes and breadfruit.
4. **Flute budding** is a form of patch budding in which the scion nearly encircles the girth of the stock stem.
5. **Ring or annular budding** is also known as girdle budding. The stock stem is completely encircled by the patch-scion.
6. **'I' budding** is similar to 'T' budding but with cross-bar incisions at both ends of the longitudinal cut.
7. **Chip-budding** is a variation of side veneer graft, but only one bud is involved.
8. **Micro-budding**, ordinary 'T' or inverted 'T' budding with very small bud.

3.2.1.2c Precautions in grafting and budding

1. Prepare root stocks for budding or grafting at least one week ahead.
2. Use only healthy materials as root stocks and as scions.
3. Select bud-wood (scion) which is in a suitable mature condition (usually about 6 to 15 months depending on the specific crop).
4. Use sharp budding knives and make all cuts cleanly and quickly.
5. Keep the time interval between incision of the root stock and insertion of the scion to the barest minimum.
6. Tie the bud firmly but not too tightly so as to avoid damage to the tender tissues.
7. Do not bud or graft during dry or very wet weather; carry out budding or grafting operations in the morning and evenings only.

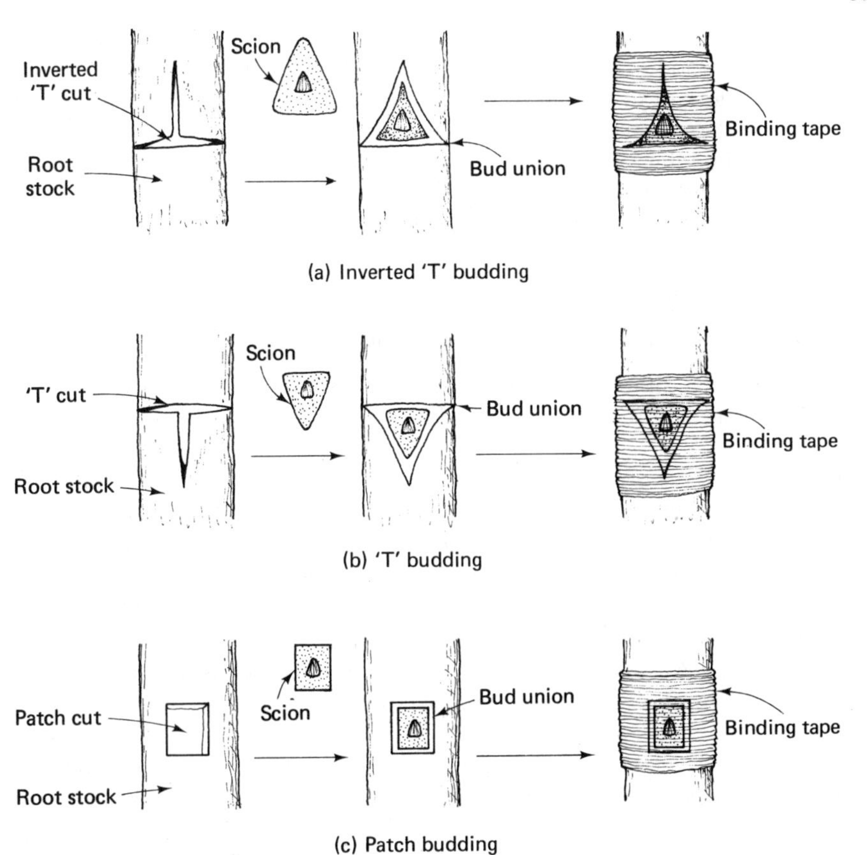

(a) Inverted 'T' budding

(b) 'T' budding

(c) Patch budding

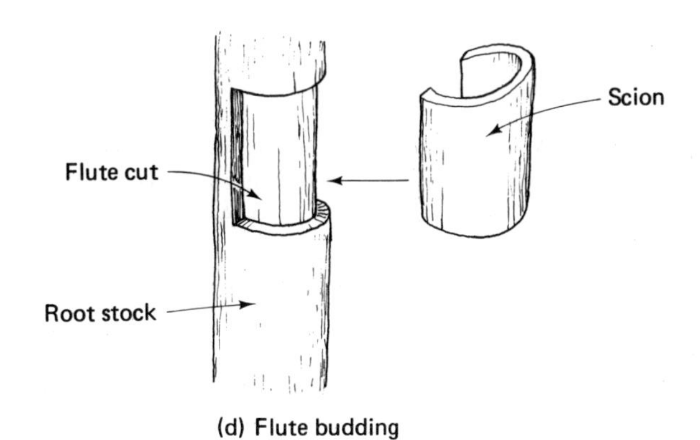

(d) Flute budding

Figure 17. Types of budding

8. Use freshly cut bud-wood. Avoid storing bud-wood as much as possible. If bud-wood is to be stored, it should be wrapped in a moist piece of sack, cloth or towel and stored in a cool place. Storage should not last for more than 24 to 36 hours before use.
9. After budding or grafting, protect the bud or graft union from rain and from direct rays of the sun.
10. Correct orientation of the scion on the stock must be ensured. The bud should not be placed upside down!

3.2.1.2d After care of grafted or budded material

Budded material should be carefully looked after. In addition, the material should be inspected for success according to the following schedule:

1. 10 to 14 days after budding: inspect budded material for indications of success. If the bud remains green, it indicates that the operation may succeed, but if the bud has turned brown and shrivelled, it shows that the budding operation has failed. If the bud remains green after 10 to 14 days, the root stock is cut back at a point about 5 cm above the bud union.
2. 25 to 30 days after budding: the successful bud will at this time start sprouting. At this stage, the stump of the root stock is cut back to about 1 cm above the bud union, paint is applied and the bud wrap is untied.

As the bud develops, the young shoot is supported with a stick; sideshoots of the root stock are removed; mulch is applied and regular watering carried out when necessary; shade is provided, fertilizer applied; pest and disease control operations are regularly carried out, if required, until the budded material is transplanted into the field. These latter provisions, of course, also apply to grafted material.

3.2.1.2e Graft and bud incompatibility

Root stock and scion may be incompatible. The earlier stock–scion incompatibility is detected, the better. There are not many reported cases of root stock–scion incompatibility in tree crops of West Africa; this does not mean that incompatibility does not occur. Where it exists, it can be easily detected by the following symptoms:

1. Failure of formation of successful graft or bud union in a high percentage of cases (when experienced people carry out the grafting or budding operation).
2. Development of a large proportion of yellow foliage in the latter part of the growing season, followed by early defoliation. This may be accompanied by a decline in vegetative growth, appearance of shoot die-back and general unthriftiness of trees.

3. Premature death of trees which may live for only a few months or years in the nursery.
4. Marked differences in growth rates or vigour of scion and root stock.
5. Overgrowth at, above or below the graft/bud union.
6. Development of swellings and/or cankerous formation around the bud union.

When incompatibility is established, it may be circumvented by the use of compatible inter-stocks or growth substances.

3.2.1.3 Layering.

Layering is a useful means of preserving plant material or even of multiplying valuable trees on a small scale. It is a time-consuming and laborious method, which may find application only in special cases. The principle is that existing shoots or branches of the tree are induced to form roots by covering part of the stem with soil (normal layering) or a moist rooting medium tied around the stem (air layering).

Vegetative propagation has been successfully applied to many tropical tree crops. Research is, however, needed in many areas including the use of growth substances to overcome graft/bud/scion incompatibility.

ADDITIONAL READING

Hartmann, T. and Kester, D. E. (1968). *Plant Propagation – Principles and Practices.* Prentice Hall, Inc., Englewood Cliffs, New Jersey, USA.

CHAPTER 4

Equipment and Materials Needed in Tree Crop Husbandry

Establishment of tree crop nurseries and plantations poses the problem of selecting the right type of equipment and materials for the required operations. This chapter intends to assist planters in the selection of their tools and materials. Equipment and materials for tree crop cultivation can be broadly divided into two categories, although the uses overlap. These are nursery equipment and materials, and field equipment and materials. Specialized tools and materials are needed for certain operations in the cultivation of individual crops, and these are listed when these crops are discussed.

4.1 NURSERY EQUIPMENT

Once the nursery site has been clear-felled and the trash carried away or burnt, it is essential to apply correct methods of cultivation and layout. For the peasant and temporary nurseries, it may not be essential to maintain a stock of tools and materials on the site. The permanent nurseries require a permanent stock of the appropriate tools and materials. The most important of these tools and materials are as follows:

1. *Cutlass or machete* This is used for cutting sticks, wood, weeds and bush undergrowth especially outside the nursery fence along the rows of the windbreak.

2. *Spade hoe or West Indian hoe* This is used for surface cultivation, preparation of bed and construction of drainage and/or irrigation channels.

3. *West African hoe* This is used for surface cultivation, weeding, earthing up of crops and clearing of irrigation and/or drainage channels and nursery paths.

4. *Spade* The spade is used generally for digging, turning over of soils (topsoil and subsoil) cleaning paths, drains, mixing soils, compost and construction of both sunken and raised beds. The spade is an indispensable tool in a nursery.

5. *Shovel* The shovel is similar in shape to the spade, but with a larger blade. Its main use is in loading soils into trailers or trolleys, sterilization tanks and in cleaning nursery paths. It can be used as a substitute for the spade in a number of nursery operations.

6. *Digging fork (3 or 4 pronged)* This is the main tool for loosening soils, burying weeds, preparing and turning compost, removing plant remains and for mixing particulate and bulky materials.

7. *Rake* This is an indispensable tool. It is used for gathering weeds after hoeing and levelling surfaces of nursery beds. It is also useful in spreading fertilizers preparatory to mixing.

8. *Handfork* This is used for lifting and transplanting of seedlings during pricking in place of a trowel. It is also used for weeding, loosening of topsoil surface, removal of weed roots and mixing of small quantities of fertilizers.

9. *Trowel* This has been termed 'horticulturist-mate'. It is a very useful tool in the nursery. It is used for transplanting seedlings, for mixing small quantities of fertilizers, soils and saw dust, for filling polybags with appropriate growing medium, for weeding, for loosening surface of soil and for application of fertilizers. The trowel is a very flexible tool.

10. *Trolleys* (*wheelbarrow*) These are essential for movement of soils, fertilizers, seedlings and other movable materials and tools around the nursery.

11. *Watering cans (10, 15, or 20 litre)* They are used for sprinkling water over the young seedlings. Different size jet nozzles should be stocked as very fine jets are needed for newly sown small seeds while medium size jets are used for growing seedlings.

12. *Spraying pumps* Spraying pumps with different nozzles should be stocked. They are used for application of fungicides, insecticides or foliar application of fertilizers.

13. *Soil sterilizer* This is a metal container for sterilizing soil. The size required will depend on the need of the nursery for sterilized soils. A soil sterilizer can be locally made from a 44 gallon (200 litre) drum by the blacksmiths. Locally made soil sterilizers from drums are less durable than imported ones. A few nursery keepers have constructed soil sterilizers from mud and these have been very durable. The only disadvantage is that the mud soil sterilizers consume more fuel than metal ones (see Figure 7). The main features of the mud-walled sterilizers are a mudwall, a metal false bottom supported with iron bars, a firing space, an emptying channel provided with a wooden plug, and a movable roof (thatch or metal). The mud-walled soil sterilizers are becoming very popular in many nurseries in West Africa.

A recent innovation in the construction of the mud-walled soil sterilizer is the replacement of the iron bar support with an ingeniously designed block wall.

14. *Pickaxe* This is used for clearing and especially for the extraction of roots from the soil.

15. *Standard axe* This is for clearing, cutting, felling trees and general use in the nursery.

16. *Knives* Three different types of knives are needed in a standard nursery. These are:

(a) Ordinary penknife – this must be very strong. Used for general purposes of cutting ropes, sharpening of pencils, etc.
(b) Pruning knife – for pruning operations.
(c) Budding knives – for budding and grafting.

17. *Secateurs* This is a scissor-like tool especially made for pruning small branches, leaves and for canopy maintenance.

18. *Headpans or buckets* These are for carrying materials and for volume measurement.

19. *Rotary hoe* This is needed in big nurseries for soil cultivation.

20. *Garden line or tape measure of 30 m length* This is needed for lining out operations and setting off right angles. It is an indispensable tool in the nursery.

21. *Measuring rod (s)* As the name implies these are rods 90 to 180 cm long, graduated in centimetres and metres for quick measurement of distances in the nursery. They are useful especially in spacing crops on nursery beds.

22. *Dibber (pointed stick)* This is needed for making planting holes for large seeds and for transplanting seedlings especially during pricking.

23. *Seed boxes* These are needed for sowing small seeds that are later to be pricked. These can be locally constructed on the nursery site.

24. *Carpentry tools* It is of advantage to keep a stock of well-maintained simple and common carpentry tools such as cutting-saws, hammers, nails, etc. in the nursery. These are often regularly needed for various purposes.

25. *Sprinklers for irrigation* A stock of sprinklers with appropriate rubber hoses will facilitate the watering of plants in the nursery.

26. *Budding tapes* These are needed for budding and grafting operations. Budding tapes have been successfully replaced with *Raphia* fibre (epidermal layer of the adaxial surface of the immature *Raphia* palm leaflets) on many occasions in different parts of West Africa. Critical experimental data are, however, needed to evaluate the use of *Raphia* fibre *vis-à-vis* the use of budding tapes.

27. *Stakes* Stakes are needed on many occasions as support for seedlings. Usually these stakes are collected from the bush adjoining the nursery or from far off. Stakes are becoming scarce especially around long-existing nurseries. The cost of collecting stakes from naturally occurring bush is becoming very high. Nowadays, it is recommended that a bush of fast growing shrubs is maintained near the nursery to serve as the source of stakes. One of the recommended shrubs is *Vernonia amygdalina*. When using *Vernonia* Stakes, care must be taken to ensure that the stakes are completely debarked and semi-dried before use, otherwise the stakes will root and become very difficult to eradicate from the nursery.

28. *Ranging poles* These are needed for laying out operations during the construction of nursery beds, paths and other structures.

Nursery operation is becoming specialized. Different types of equipment are being invented to meet the needs of specialized nurseries. The nurseryman must constantly keep abreast of the equipment so as to be up to date in the selection of his equipment.

4.2 NURSERY MATERIALS

Many different types of materials are needed in the nursery. This will depend on the specific crop being raised. However, the following materials are often needed in a nursery:

1. *Polybags* These are obtainable from the manufacturers. When labour was cheap, cane baskets were preferred to the polybags. Nowadays, the cane basket is too expensive and for economy, polybags are in common use.

2. *Growth substances* IAA (indole acetic acid), NAA (naphthalene acetic acid), and gibberellins are needed for rooting cuttings and for inducing early germination in dormant seeds.

3. *Fertile topsoil* This is one of the prerequisites for selecting a site for a nursery. Where nursery beds are used for sowing and raising of seedlings, the topsoil of the nursery site should be fertile, or fertility has to be raised with the aid of admixtures as farmyard manure or compost. When seeds are sown and raised in polybags, cane baskets or seed boxes, the topsoil of the nursery may not be fertile. Fertile topsoil can then be collected from elsewhere. During the establishment of nurseries where plants are exclusively to be raised in containers, it is of advantage to scrape and preserve any fertile topsoil for later use. This saves the cost of transporting topsoil from outside the nursery site.

4. *Sawdust* Generally, germinating seedlings require light, fertile soils. To lighten the texture of topsoils for good growth of seedlings, it is an advantage to maintain a regular supply of well-settled sawdust in the nursery.

5. *Mulching materials* Seedlings of a number of West African tree crops require mulching in the nursery prior to transplanting. It is an advantage to maintain a source of mulching material close to the nursery.

6. *Insecticides and fungicides* A stock of different types of insecticides and fungicides must be maintained. As plants are crowded together in the nursery, each disease or pest outbreak must be promptly treated.

7. *Herbicides* These are used for weed control particularly in controlling weeds along nursery paths and boundaries.

8. *Manures and fertilizers* Manures and fertilizers are not normally needed in the nursery provided fertile topsoil is available. Nevertheless, it is desirable to keep a small stock of complete fertilizer mixtures preferably those carrying traces of micronutrients.

9. *Planks* It is customary to keep a stock of various sizes of wood (planks) in the nursery. They are needed for many varied constructions such as seed boxes and seed trays.

The materials needed in nurseries will vary with the nature and intensity of operations. The list above is not exhaustive. The golden rule is to stock the nursery store with whatever is needed well ahead of the time of use.

4.3 FIELD EQUIPMENT

For planting, maintenance and harvesting of tree crops in the field, various tools are needed. While nursery tools are more or less common for most of the crops, field tools are more specialized. The specialized ones are treated under each crop while the more common ones are listed below:

1. *Bush clearing equipment* The type of tool or machine used for bush clearing will vary with the size of plantation. For big plantations, the bush is cleared with heavy machinery such as bulldozers and treedozers. For small farms, the bush clearing tools consist of machetes, axes, pickaxes, mattocks, sharpening stones and files.

2. *Lining up equipment* Lining up is a survey operation. The amount of surveying to be done will depend on the nature of the land. Sloping lands will require contouring either for terracing or for contour planting, in which case levelling equipment such as dumpy levels are needed. On fairly level lands, levelling may not be necessary in which case the lining tools will consist of machetes, ranging poles, garden lines, pegs, mallets and steel measuring tapes. The lining up operation is discussed in Chapter 5.

3. *Digging equipment* Generally, holes for transplanting seedlings of tree crops in West Africa are dug by hand. The tools required are the machete and the shovel. With increasing shortage and high cost of labour, it is desirable to machanize this operation, especially on large plantations. Tractor operated post-hole diggers may be particularly useful.

4. *Planting or transplanting equipment* The equipment required for transplanting consists of a sharp knife for cutting off polybags, and shovels or spades for filling planting holes with soil.

5. *Field maintenance equipment* Requirements for various tools in the maintenance of tree crop farms in West Africa are gradually changing. One direction is towards increasing mechanization of the maintenance operations of tree crops when cover crops are not necessary or when they have failed to establish. The second one is towards more efficient use of manual operations when mechanization proves difficult. A detailed list of required equipment for maintenance is, therefore, difficult and unrealistic. Nevertheless, the following tools are indispensable: machetes, weeding hoes, extension ladders, pruning knives, lawn mowers (either hand operated or tractor mounted), spraying pumps (various types and sizes), recording materials, harvesting knives, processing, grading and packing tools (discussed under each crop).

Sometimes soil augers are required to provide an opportunity to sample the soils in the plantation to allow chemical analysis. Soil laboratories can provide valuable advice on fertilization practices on the basis of correctly collected soil samples.

As farm labour is becoming scarce in the tree crop growing areas of West Africa research activities are being concentrated on how best to mechanize maintenance, harvesting, processing and grading operations. Appropriate tools will be recommended by researchers from time to time and palnters should keep abreast of any new developments.

4.4 FIELD MATERIALS

The exact materials needed for field operations will vary with the individual tree crops. Nevertheless, it is advisable for planters or farm managers to keep a stock of the right type of fertilizers, plant protection chemicals, herbicides, growth regulators and mulching materials. Mulching needs to be carried out yearly until the young trees close their canopy. After the canopy has been closed, there may be accidents such as tornadoes, fallen trees and pest damage that could cause a break in the canopy. Soil below such breaks must be mulched in addition to other steps that are taken to repair the damage.

As already pointed out, a considerable effort is being put into research work on how best to maintain tree crop fields. Undoubtedly, the results of work in progress will develop recommendations on novel means. The manager or the planter must, therefore, constantly bring himself up to date with the literature.

CHAPTER 5

Planting in the Field

5.1 SELECTION OF THE PLANTING SITE

The main factors that determine the suitability of a site for growing any tree crops are as follows:

1. *Altitude* Generally tree crops in West Africa are low-altitude crops. With the exception of *Coffea arabica*, most are cultivated from sea level up to an altitude of 700 m.

2. *Topography* Level sites are most suitable for growing tree crops, because they facilitate farm operations, movement of men and materials, and reduce the cost of anti-erosion measures such as terracing.

3. *Drainage* Soils that are well drained are the most suitable for most tree crops. However, breadfruit, jakfruit and *Raphia* palm prefer very wet, but not waterlogged, sites.

4. *Aspect* Strong winds are harmful to all tree crops. Most harmful is the desiccating harmattan wind which prevails over a considerable part of West

Africa during the dry season. In selecting a site, the direction and duration of the prevailing winds must be considered.

5. *Climatic factors* Most tree crops in West Africa thrive best under humid tropical conditions. Therefore, the major climatic factors that must be considered in selecting a site are the following:

(a) Temperature – an annual average temperature of 25 °C to 29 °C is ideal for most tree crops in West Africa.

(b) Rainfall – when considering rainfall it must be noted that the tree crops of West Africa are predominantly evergreen, in a few cases semi-deciduous. Therefore, they require a satisfactory supply of moisture all the year round. Most of these tree crops are adapted to an annual rainfall of 1000 mm and above. Few, however, like cashew, mango, guava and date palm can thrive in low rainfall areas (< 1000 mm). More important than the amount of rainfall is its distribution. The available rainfall should be distributed as evenly as possible over at least nine months of the year.

(c) Insolation – tree crops of West Africa require much sunshine. Growth becomes poor in areas of frequent and dense cloud covers. Yield reductions of over 30 per cent in cacao have been attributed to cloudiness in some parts of West Africa (Opeke, 1965).

6. *Soils* Permanent crops, as tree crops, occupy the same site all through their life and their performance over the years is related to the characteristics of the soil in that spot. Therefore, the soils should be deep, fertile, well drained, free from soil-borne pathogens and supply sufficient water, air and nutrients.

7. *Labour supply* The culture of most tree crops of West Africa is labour intensive. Many of the operations have not been mechanized. It is also yet to be seen whether a number of operations can be economically mechanized. Therefore, in selecting a site for any of the crops, the source of labour must be given serious consideration.

8. *Accessibility* Accessibility of the site is of great importance in siting tree crop plantations. Accessibility by road, rail, water and/or air affects the supply of services – labour, agricultural chemicals, machines, advisory services and marketing of farm produce. Accessibility is a procurable asset, but it must be taken into consideration in site selection.

All these factors are important in selecting sites for tree crop cultivation. It is, however, rarely possible to obtain an optimum combination of all the factors in a single location, and the prospective planter is expected to make a judicious choice of the best combination.

5.2 PREPARATION OF THE PLANTING SITE

Most tree crops in West Aftica in the past were established by establishment into food farms. With the increasing importance of cacao and oil palm produce in the world market, the system of monocropping was adopted for these crops in the early 1930s, and this led to land preparation specifically for planting these tree crops. There are three standard methods of land preparation for this purpose:

1. *Clear-felling* The area is underbrushed; all the trees, small and big, are felled; the trash is either burnt or carried away from the site. The site is cleared of all vegetation. Experience has shown, however, that soil deterioration on clear-felded land is very rapid. To reduce the fast rate of soil deterioration when clear-felling is adopted, it is advisable to leave as much as possible of the leaf litter and felled vegetation on the site. This has its own danger of encouraging the growth of pathogenic root fungi. It is also advisable to plant leguminous cover crops such as *Centrosema pubescens*, *Pueraria phaseoloides*, *Calopogonium mucunoides* or *Stizolobium deeringianum*. When the crop to be planted requires shade trees, it is important to establish these immediately upon clearing the land.

2. *Selective thinning* Selective thinning of forest provides a means of establishing particularly those tree crops of West Africa that require shade during early growth – cacao, kola, coffee, breadfruit, breadnut, etc. It is also a cheaper method than clear-felling. The site is underbrushed, followed by felling the big trees which constitute the upper storey. If the under storey provides more shade than necessary, it is thinned to provide a light, reasonable even shade over the site: then traces are cut at the selected spacing, holes dug and the seeds, seedlings, or rooted cuttings planted. The litter and trash remaining decompose and release nutrients and increase organic matter in the soil.

3. *Bush fallow traces or simple felling* This is clearing of land under bush fallow for planting of tree crops. This is the simplest type of land preparation, but it can be applied only if the original upper storey and the under storey trees have already been felled during former cropping cycles. It consists of cutting traces at the selected spacing through the fallow bush, digging the planting holes along the traces, planting the crop. The remaining bushes are cut back gradually as the crop develops its canopy in subsequent years.

Table 4 shows the relative costs of establishment operations under the three standard methods of land preparation (data from Cocoa Research Institute of Nigeria, Ibadan).

The relative rates of growth of kola plants established on land prepared according to the three standard methods are shown in Table 5.

5.3 LINING OUT OF THE FIELD

After land clearing, the next important operation in the preparation of the planting site is the lining out of the field. On small farms, i.e. farms which are less than 5 to 10 ha, the lining out operation consists of locating planting sites on the field. On large farms, there are two aspects to the lining out operation:

Table 4 Relative cost of various operations in three differently established kola plantations, Ibadan. Total costs of clear-felled plantations per hectare over the first three years was taken as 100 (Copied from van Eijnatten, 1969)

		Clear-felling (2.3 ha)		Simple felling (6.0 ha)		Selective thinning (3.9 ha)	
Preparation of the land (underbrushing, packing, uprooting, cutting of traces)		52		42		11	
Holing and planting		4		6		5	
Weed and shade control	1st year	7	28	9	28	4	20
	2nd year	9		7		8	
	3rd year	12		12		8	
Mulching	1st year	8	16	6	12	4	08
	2nd year	8		6		4	
Total costs		100		88		44	

Table 5 Relative growth in height (cm) of kola seedlings established in three different ways (data from van Eijnatten, 1969)

Number of years after establishment in the field	Methods of establishment: Clear-felling	Simple felling	Selective thinning
	Seedlings		
1	82(± 5)	76(± 5)	79(± 2)
2	134(± 5)	148(± 10)	131(± 3)
3	182(+ 6)	—	—
	Ramets		
1	60(± 4)	63(± 2)	57(± 4)
2	90(± 4)	133(± 3)	93(± 6)
3	137(± 5)	—	—

1. *Blocking of the plantation* This is the process by which the plantation is divided into convenient, sizeable blocks; each block being 5 to 10 ha in size depending on the spacing of the crop. For closely spaced crops, the block size is nearer to 5 ha while large blocks are preferred for widely spaced tree crops.

The blocks are separated from one another by 4 m wide farm roads. The shape of the block should normally be rectangular. It may be a rectangle of four equal sides, i.e. a square, but experience has shown that rectangular blocks of length 1.5 times the width, facilitate estate operations much more than square blocks.

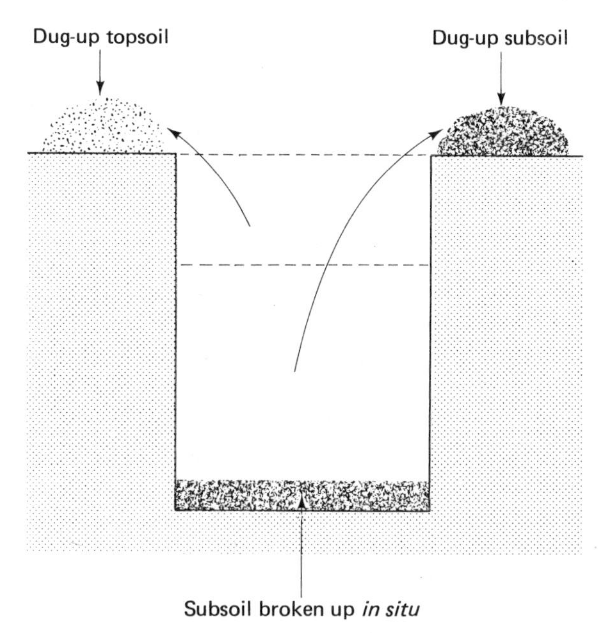

Figure 18. A typical planting hole

2. *Locating the planting sites* After division of the plantation into blocks, the planting sites are to be located. Materials required for this purpose are ranging poles, measuring chains or tapes (30 m) and pegs or stakes to mark the planting site.

A baseline is adopted at one side of the block in which the planting sites are to be marked. Along this baseline the various sites are marked at the appropriate distances required for the crop to be planted. If planting is done 'on the square' a right angle is to be constructed on the baseline in order to locate further sets of planting sites. This is done using the 3-4-5 principle. All planting sites, once located, are marked with wooden pegs pending the preparation of the planting holes. Preparation of the planting holes is done after lining out of the field. Common dimensions for the planting holes are 60 × 60 × 60 cm. During digging, topsoil is preserved at one side of the planting hole and subsoil at the other side. Once the required depth of 60 cm has been reached, the subsoil at the base of the hole is loosened (see Figure 18).

5.4 PLANTING AND TRANSPLANTING OPERATIONS

Ideally, seedlings are uprooted from the nursery with a ball of earth on the morning of the day of transplanting. Where seedlings are raised in field nursery beds and are to be transported over considerable distances for planting, they are usually uprooted a day or two ahead of planting, treated with clay slurry, transported during the cool hours of the evening or night to the planting site, rested and planted as early as possible, preferably on the day of arrival at the planting site.

Seedlings raised in polybags or cane baskets are transported to their planting sites without many problems. With such seedlings, a rest period of one to two weeks must be allowed on arrival at the planting site before planting.

Transplanting is a two-man operation. The operations are in the following sequence:

1. The planting hole is half-filled with topsoil mixed with rotten compost or topsoil with high organic matter content.
2. The seedling is put in position with one operator holding it upright and in position in the hole, and the other operator being responsible for arranging the roots properly in the hole (for seedlings transplanted with naked roots) or removing the polybag, and filling the remaining portion of the hole with topsoil. After proper filling, the second operator consolidates the soil around the seedlings to ensure that air pockets are completely excluded from the root zone (see Figure 19).

Transplanting is a very delicate and extremely important operation. Transplanters should be selected on the basis of carefulness and be trained in the correct ways of planting.

Efficiency of transplanting is evident when the following aspects are satisfied:

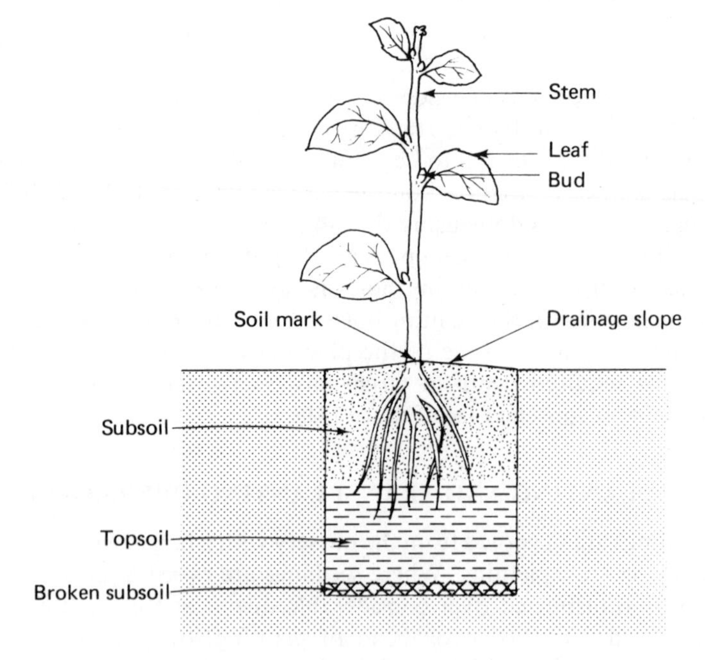

Figure 19. Transplanted seedling

1. Drainage of rain water away from the seedling should be allowed for by the sloping surface of the topsoil replaced in the planting hole.
2. The transplanted seedling should be planted in such a way that the original soil mark is level with the surface of the soil.
3. The seedling must be firmly embedded in the topsoil. Pulling on the top leaf should not dislocate the seedling.

5.5 SPACING

Adequate planting distances are essential for economic productivity of tree crops. Table 6 shows the methods of propagation, average age at first bearing and usual spacings for most of these tree crops.

5.6 POST-PLANTING MAINTENANCE OPERATIONS

The maintenance of tree crop farms is an all-year-round activity. It embraces various operations at different periods of the year. Maintenance operations vary with such factors as climate, soil fertility, intensity and distribution of rainfall, age and vigour of plantations and method of establishment. Adequate field maintenance helps to reduce the risks of weed, disease and pest infestations. The major items of post-planting maintenance are weeding, mulching, shade management (where shade is required), pruning and removal of lateral suckers, replacement of dead stands, pest and disease control.

5.6.1 Weeding

This is the removal of unwanted plants and shrubs from the plantation. Weeding is a great problem when fields are badly established, especially where cover crops are not properly established and, generally, the weed problem is greatest during the first four to five years, before the plants have established a closed canopy. Frequency of weeding, type of weeding and use of weedkillers will vary with the type of crop. General guidelines are as follows:

1. 'Clean-weed' the tree crop farm. For plantations, row weeding thrice a year supplemented with at least four to six slashings of the avenues per year is recommended.
2. Mulch the trees towards the end of the rains. Grass mulch is recommended. Avoid placing trash close to the base of the plants.
3. Plantain is recommended as a good nurse shade. Bananas should be avoided as nurse shade. Any thinning or the final removal of nurse shade should be done at the end of the dry season.
4. Supply of vacancies should be done during the first two rainy seasons after the original establishment of the new farm. This should be done from the time that the early rains are steady.
5. Prune trees regularly to maintain good shape. Canopy maintenance and

Table 6 Basic data on the commonly grown tree crops of West Africa

Name	Methods of propagation	Average age at first bearing (years)		Recommended spacing (m)
		Seedlings	Vegetatively propagated plants	
Cacao	Seeds Vegetative budding Rooting Marcotting Grafting	4	2–3	3.1 × 3.1
Cola nitida	Seeds Vegetative budding Rooting Grafting Marcotting	5	3	6.1 × 6.1
Cola acuminata	Seeds Vegetative budding Grafting Marcotting	7	4–5	6.1 × 6.1
Coffee	Seeds Vegetative budding Layering Grafting Marcotting Rooting	4	3	3.1 × 3.1
Rubber	Seeds Vegetative budding Layering Grafting Marcotting Rooting	5	4	4.8 × 4.8
Cashew	Seeds (vegetative-experimental)	4	3	4.8 × 4.8
Avocado pear	Seeds Vegetative budding Rooting	7	5	8.1 × 8.1
Thea (Tea)	Seeds Vegetative budding Rooting Marcotting	3	2	2.5 × 2.5
Guava	Seeds Rooting Budding Layering Marcotting	5	3	6.1 × 6.1
Mango	Seeds Rooting Budding Grafting	4	3	12.0 × 12.0

Table 6 (*Continued*)

Oil palm	Seeds only	4	—	10.0 Triangular
Coconut	Seeds	5	—	8.1 × 8.1
Raphia palm	Seeds	—	—	8.1 × 8.1
Akee apple	Seeds	7	—	6.1 × 6.1
Sweet orange	Seeds Vegetative budding	4	3	7.0 × 7.0
Sour orange	Seeds Vegetative budding	4	3	7.0 × 7.0
Lime	Seeds Vegetative budding	6	4	6.4 × 6.4
Lemon	Seeds Vegetative budding	5	4	7.0 × 7.0
King orange	Seeds Vegetative budding	4	3	7.0 × 7.0
Grapefruit	Seeds Vegetative budding	5	4	8.1 × 8.1
Shaddock (pomelo)	Seeds Vegetative budding	6	4	8.1 × 8.1
Sweet lime	Seeds Vegetative budding	7	5	6.0 × 6.0
Tangerine Mandarin	Seeds Vegetative budding	6	3	7.0 × 7.0
Satsuma Orange Tangelo	Budding	—	4	7.0 × 7.0
Breadfruit	Root cuttings	—	5	11.0 × 11.0
Breadnut	Seeds Vegetative-patch-budding	6	4	11.0 × 11.0
Jakfruit	Root cuttings	—	4	11.0 × 11.0
Sweet sop	Seeds Vegetative-patch-budding	6	4	6.1 × 6.1
Sour sop	Seeds Vegetative-patch-budding	6	4	6.1 × 6.1

light pruning may be done at any time of the year. Heavy 'formation' pruning should be done in May–June. Pruning equipment should be sharp. Pruned surfaces should be protected with coal tar, tree sealing compound or ordinary paint.

6. Remove all disease infected plants immediately they are recognized.

5.6.2 Mulching

This is the provision of a surface layer of dead vegetable matter to keep the soil around the plants moist and enrich it with organic matter. The application of mulch is a cultural method which yields handsome dividends when properly carried out. Mulch is beneficial to crop plants in five important ways:

1. Mulch conserves moisture round the plant.
2. It helps to cool the soil around the plant.
3. It adds organic matter to the soil on decomposition.
4. It helps to suppress weeds round the plants.
5. It acts as an anti-erosion measure.

When the advantages of mulching are added to those of other maintenance operations, growth is superior and this results in more vigour, early canopy formation and early heavy yields.

The plant should be provided with mulch, 15 cm thick and covering an area from 15 cm to 1 m away from the base of the plant. The distance of 15 cm away from the base of the plant is to discourage termites which might first attack the mulching material and then migrate to the seedling. Vegetative mulch is preferred to synthetic mulch such as polythene sheets, first because of high costs and second because the polysheet does not add nutrients to the soil.

5.6.3 Shade Management

Many tree crops in West Africa originated as part of the under storey of the tropical forest and require shade during their first few years after planting in the field. The ideal shade plant should satisfy the following conditions:

1. It should be easy to establish.
2. It should be able to provide an evenly distributed shade within a short time of planting and throughout the year, especially during the dry season when shade is most needed.
3. It should compete as little as possible with the crops for soil nutrients and especially for moisture.
4. It should be easy to remove when no longer needed, and the process of its removal should do as little damage as possible to the crop.
5. It should not be an alternative host to insect pests and diseases of the crop.
6. If possible, it should have a commercial value.

So far the ideal shade plant has not been identified, but plantain is nearest to an acceptable shade plant. As has been shown above, natural shade is used, land preparation is done by 'simple felling' and/or by selective thinning.

In addition to permanent shade (i.e. shade that lasts till complete canopy formation by the crop), nurse shade is recommended for a few tree crops of West Africa. Cocoyams (*Colocasia esculentum*, *Xanthosoma sagitifolium*) have been successfully used for cacao in Ghana and coffee in Ivory Coast. Nurse shade should be removed after two or three years.

In addition to various shade plants, temporary shade may be needed. This is generally necessary during the first dry seasons after planting into the field. Temporary shade can be provided through the use of palm fronds (see Figure 20). The temporary shade is removed at the end of the dry season, i.e. when the early rains become steady.

Figure 20. Temporary palm frond shade

5.6.4 Pruning

Pruning is carried out, essentially, to remove unwanted growth and thus maintain regularly shaped trees carrying a well-balanced, firmly closed canopy. Pruning is sometimes done with a view to regenerate old trees. Unskilful pruning may lead to unnecessary injuries to trees with consequent reduction in yield. Light pruning is recommended although the frequency of pruning depends on the rate of growth of trees. Unwanted vegetative parts (lateral suckers, chupons) should be removed as close as possible to the stem from which they emerge, preferably at an early stage, to save the trees from wasting their energy on unwanted growth.

Canopy maintenance, the removal of unwanted growth from the canopy and light pruning can be carried out at any time of the year. Heavy pruning (**formation pruning**) which is aimed at regulating the shape of the trees should only be carried out when the early rains become steady. A lightweight sharp cutlass or pruning saw are the most suitable tools for most pruning operations. A sharp knife or secateurs should be used for removing chupons, lateral suckers, young flushes and major branches. A heavy bowsaw is recommended for removing big branches and stems. Pruned surfaces should be painted with ordinary paint or any sealing compound. Lateral suckers are generally profuse and troublesome in citrus and coffee. Their prevalence in budded citrus, especially around the bud union, is indicative of incompatibility between root stock and scion.

5.6.5 Disease and Pest Control

It should be borne in mind that when a high standard of sanitation is adhered to in tree crop plantations, incidence of disease and pests is greatly reduced. Nevertheless, very many diseases and pests may attack the tree crops of West Africa.

The most dangerous group of diseases is that of the viruses such as the swollen shoot virus disease of cacao (SSVD) and the tristeza of citrus. Equally important are the pod rot diseases of cacao, brown rot gummosis of citrus, panel disease of rubber and bud rot disease of coconut all caused by fungi belonging to the *Phytophthora* group.

These diseases and their methods of control are outlined under each crop. However, efficient control of diseases and pests of crops in West Africa will, in the future, only be achieved through an integrated approach, which is a judicious use of natural control measures (plant resistance to pests/diseases, parasites of pest or disease) prior to deciding on hazardous chemical control measures.

5.6.6 Establishment and Management of Cover Crops

Leguminous plants such as *Centrosema pubescens* are often planted as cover crops in tree crop plantations in West Africa. The use of a cover crop is more popular with those tree crops that are widely spaced – kola, oil palm cashew, rubber, citrus and so on. Recently, scientists at the IFCC, Cote D'Ivoire, demonstrated that cover crops can be used even with unshaded robusta coffee spaced at 3 × 3 m. The main advantages of leguminous cover crops in tree crop plantations are as follows:

1. The cover crop serves as an anti-erosion measure.
2. It helps to suppress weed.
3. It adds nitrogen to the soil through its nodules.
4. It adds organic matter to the soil.
5. It helps to cool the soil.
6. The roots help to improve the physical structure of the soil.

Cover crops should be established as early as possible after land preparation. Where land is clear-felled, it is advisable to establish cover crops before planting the main crop. When simple felling and/or selective thinning is practised, the cover crop can be established after field planting.

The management of the cover crop is as important as its establishment. Cover crops should be kept away from the crops by a space of at least 30 to 40 cm around each plant. As the name implies, the cover crop is to cover the interspaces between the plants. Where a cover crop is used it is not advisable to use non-selective herbicides. Improperly managed cover crops can turn out to be the most dangerous weed in the plantation!

5.6.7 Watering/Irrigation

In the ideal tree crop area such as the Ikom area of Nigeria, the southern Ashanti area of Ghana, Divo in Ivory Coast, watering and/or irrigation of tree crops in the field is not necessary. These areas have adequate rainfall distributed throughout the year. In many parts of West Africa, the dry season may last for up to three months. In such areas, success in the establishment of tree crops in the field will depend to a great extent on the supply of water to the seedlings during their first two dry seasons in the field. Adequate mulching and establishment of cover crops helps to economize on the application of water. Watering should be carried out during the cool hours of the day, preferably in the mornings or the evenings, and water should not be allowed to collect around the seedlings.

5.6.8 Replacement of Dead Trees

Supplying of vacancies is usually done during the first two rainy seasons after field planting of the new farm. This will ensure a measure of uniformity among the trees. Supplying operations should be done as soon as the rains become regular. Variation in climatic conditions makes the recommendation of a specific time difficult, but it should be noted that the essential point is adequate rainfall during and after the operation. Supplying operations carried out later in the rainy season often allow the seedlings a short period of rainfall for growth before the onset of dry weather which may result in heavy casualties. It is relevant to mention that the earlier the vacancies can be filled, the better are the chances of these seedlings catching up to give a uniform tree crop plantation. If supplementary irrigation can be provided, the filling of vacancies becomes less dependent on climatic factors.

5.7 CROPPING PATTERNS WITH TREE CROPS

Where land is abundantly available, the normal practice with tree crops is monocropping. However, land is gradually becoming limited in most tree crop producing countries of West Africa. Also, many tree crops of West Africa are widely spaced, and in the early years before these crops form their canopy, it has been shown that intercropping with compatible annual crops is an economic advantage.

When intercropping, a space of 1 m must be left to the tree crop seedling in the first year. In subsequent years, this distance is gradually increased until intercropping is stopped when canopies close in.

It is essential to adopt adequate fertilizer practices for the intercrop in addition to those for the tree crop.

Another method of cropping adopted for tree crops in the drier parts of West Africa is the '**compound cropping system**'. In this system, a few trees are grown around the compound and they are supplied with their moisture requirements from household waste water. Compound crops also receive a large part of their nutrients from household wastes – ashes, food remains, etc. Generally, compound crops receive more intensive care and they are generally more productive than field grown crops. However, they are necessarily limited in number.

ADDITIONAL READING

Opeke, L. K. (1965). The biological factor of light intensity on cacao. *Proc. 1st International Cacao Research Conference*, Abidjan, Ivory Coast.

Soyele, W. A. and Bolaji, E. (1971). *Progress in Tree Crop Research in Nigeria* (*Cacao, Cola and Coffee*). Published by the Cocoa Research Institute of Nigeria, pp. 68–77.

CHAPTER 6

Regeneration and Rehabilitation of Tree Crops

Regeneration and rehabilitation are two terms which are often confused in tree crop literature. Regeneration refers to regrowth of damaged trees. A tree can be damaged by storms, animals, falling trees, etc. The recovery of such damaged trees is known as regeneration. Rehabilitation is the act of renewing a tree crop farm that had become derelict either because of old age or because of disease and/or pest infection.

Power of regeneration may be one of the characteristics of the crop used in achieving rehabilitation of the farm!

6.1 REGENERATION

During storms heavy trees may be blown over or badly damaged by breakage. However, dormant buds soon become active and develop rapidly. Several regularly spaced vigorous chupons at the base of the trunk should be retained, but probably only few chupons should eventually be allowed to grow up. Thus a well-balanced canopy will develop. Earth is heaped up around the base of the new shoot to encourage growth of an independent root system. Chupons growing closest to the ground are, therefore, preferred to those growing higher up as it is easier to earth them up.

The most difficult part of the operation is to fit the young plant into the overall canopy. In most cases, the new tree will flower and bear fruit much earlier than newly planted seedlings. The surrounding trees should be prevented from 'choking up' the new tree by regular removal of their growth into the gap, originating from the disappearance of the original tree. The control of regeneration is an important operation in order to maintain continuity of yields from a tree crop farm.

There are other types of common damage that can occur in a tree crop farm.

1. *The apical bud of a seedling may by damaged* In this case, two strong buds on the remaining stem of the seedling are allowed to develop. The two shoots are nursed but only that shoot which branches first is retained for canopy formation.

2. *The canopy of a tree may be badly damaged* The canopy of a tree can be damaged in a number of ways. A main branch can be badly torn through the branching base. In this case the only repair possible is to trim the broken surface, paint it and allow the other branches to continue growth.

A branch may be broken while the tree still carries over half of the canopy. In this case, the broken branch in removed, and the stump allowed to regenerate. Trimming must be carried out as necessary. Damage to smaller branches in the canopy is treated by removal of the damaged branches.

3. *A tree may fall over and be badly damaged* If the damaged tree is a vigorous one, regeneration should be allowed. In case of average or poorly growing trees, the stump should be removed and a new seedling planted in its position. A fast growing shade plant which can subsequently be easily removed could be established close to the gap in the canopy temporarily and thus prevent adjacent trees from shading the new crop plant. As root competition between the newly planted seedling and the surrounding established trees may retard the growth of the seedling, roots of the adjacent crop plants should be cut. Similarly, branches from old trees that may over-shade the newly planted seedlings should be pruned.

6.2 REHABILITATION

Rehabilitation of tree crop farms may be necessitated by several factors. The following are the most common ones:

1. Old age of the plantation.
2. Infestation by diseases, especially virus diseases.
3. Necessity to change the variety of trees in the old plantation to a more productive variety.

The methods of rehabilitation adopted will depend on the cause which necessitated rehabilitation.

When rehabilitation is necessitated by old age, or by the need to change to a new variety, the following methods of rehabilitation are recommended.

1. *Coppicing* After coppicing the stumps are allowed to develop two or three strong shoots. The farm is clean-weeded and appropriate fertilizer applied. The best two of the three shoots are retained and allowed to develop mature canopies.

Alternatively, the two shoots could be used as root stocks onto which is budded or grafted the scion of the desired new varieties. This is top-working.

This method has been successfully applied to cacao in Nigeria (Are and Jacob, 1970).

2. *Interplanting the old trees with young seedlings* When young seedlings are planted through the old ones at the desired spacing, the old trees are used as shade for the young ones. This method is extensively used with cacao. Results obtained have shown that young cacao performed best in growth and yield when old cacao trees were kept as shade when compared to those obtained by complete replanting. The use of old trees as shade over young trees has the following advantages:

(a) It saves the planter the investment on the establishment of alternative shade.
(b) It enables the planter to continue harvesting the old trees, thus preventing the occurrence of a sudden break in the income flow of the planter.
(c) It helps to suppress weed, thus reducing the cost of farm maintenance.

This methods also has disadvantages. Some of these are as follows:

(a) The fact that the old trees may act as a reservoir for diseases and pests.
(b) The old trees, taking advantage of added fertilizer, may over-shade the young ones and thereby depress their growth, unless properly managed.
(c) When the old trees are to be removed, much care is needed to minimize damage to the young trees.

3. *Complete replanting* Complete replanting is the most widely used method of rehabilitating tree crop farms, especially where rehabilitation is necessitated by disease infection. In this case, old trees are completely removed from the plot (complete removal is essential especially when the old trees are disease, especially virus, infested). The field is lined out, pegged, holes dug, new seedlings planted, cover crop and shade plants established.

Complete replanting permits orderly replanting and is recommended for disease and pest infested farms. The method is expensive; it encourages rapid weed growth, increases soil erosion and pest infestation. For healthy tree crop farms, or where there is need to change the variety of a tree crop, '**phased replanting**' is recommended. In this case, the rehabilitation is phased over a number of years and it enables the planter to earn some income during the period that the farm is being rehabilitated.

Three factors call for further developing methods of rehabilitating tree crop farms other than the complete replanting method. First is the problem that farmers are opposed to cutting out trees because they regard the trees as a legacy to be passed on to their children. Second, farmers do not want to invest in either seedlings or nurseries, because these are expensive. Third, holdings of peasant farmers are small because of limits set by inadequate finances, family labour, and the operation of private land ownership systems.

ADDITIONAL READING

Are, L. A. and Jacob, V. J. (1970). Rehabilitation of cacao with chupons from coppiced trees. *Cacao* (*Costa Rica*), **15** (1), 1–4.

Odegbaro, O. A. (1975). Prospects of rehabilitating Amelonado cacao with improved cacao varieties in Nigeria without completely replanting. *Proc. 5th International Cacao Research Conference*, Ibadan, Nigeria.

Odegbaro, O. A. and Folarin, J. O. (1974). Potential pod production and pod yields of F3 Amazon cacao budded or grafted onto shoots regenerated from coppiced Amelonado trees. *Turiabla*, **24** (3), 256–264.

CHAPTER 7

Cacao

Cacao, *Theobroma cacao* L., is of great importance as a commodity in the world trade.

7.1 HISTORY

7.1.1 The Origin of Cacao

Cacao developed in the Upper Amazon region of Latin America. In classical Mexican mythology, cacao, one of the foods of the gods, originated in the Garden of Life. When mortal man sinned and was driven out of the Garden of Life, the gods in compassion sent a number of food crops to man in his exile. Quetzalcoatl, the god of air, took the seeds of quachahuate – the cacao, tree – to man and stayed with him for a while in exile, growing cacao, corn and other food crops. Man nicknamed Quetzalcoatl as the 'Garden Prophet'. The story further goes, that due to envy, a principal god forced Quetzalcoatl to flee. In remembrance, man then erected altars and burnt incense to the 'Garden Prophet', under the name VOTAN, hoping that one day the beloved 'Garden Prophet' would return. Meanwhile, the cacao was considered to be man's inheritance from Quetzalcoatl.

In Mexico where cacao was first discovered and grown, inclusion of cacao beans in religious rituals and performing religious rites in the husbandry of cacao were all based on this mysterious origin of the crop.

The word cacao in modern usage refers to the tree while the word cocoa refers to a drink made from its seeds. The words 'cacao' and 'chocolate' arose from the Mayan and Aztec languages. The Mayans and the Aztecs were recorded as the first to use cacao.

Extensive variability of naturally occurring cacao trees exists in the region of the Upper Amazon Basin. De Candolle in his *Origin of Cultivated Plants* stated that cacao has been cultivated in America for 3000 to 4000 years. Until the beginning of the sixteenth century no mention of cacao is traceable in the literature of any other nation except the Latin American countries of Brazil and Mexico. The crop was discovered by Christopher Columbus during his fourth voyage to the new world. Columbus intercepted a canoe loaded with various agricultural products including cacao, off the coast of Yucatan, and took a sample to Spain. The specific centre of origin of cacao has been accepted as the area from the forests of the Amazon to Orinoco and Tabasco in southern Mexico.

7.1.2 Introduction of Cacao to West Africa

When Cortes conquered Mexico in the sixteenth century, he carried cacao beans back to Spain. Wealthy Spaniards took kindly to cocoa drinks and the habit of drinking cocoa spread rapidly to Italy, France, Germany, the Netherlands, and later to other parts of Europe (Opeke,1969). To sustain an interest in cocoa drinks and to obtain regular supplies of cacao beans at low prices from its colonies, Spain introduced cacao to Africa. Although relevant documents relating to the introduction of cacao to Africa are yet to be assembled, it is generally believed that cacao was introduced into Fernando Po around 1840 by William Pratt.

Cacao was introduced into Nigeria from Fernando Po by Chief Squiss Ibanningo in 1874 at approximately the same time that Teteh Quashi introduced the crop into Ghana. Other sources of introduction of the crop to West Africa include trading companies, Christian missionaries, soldiers, chiefs, farmers' associations, cooperatives, the various Departments of Agriculture and more recently the West African Cocoa Research Institute (WACRI), the Cocoa Research Institute of Ghana (CRIG), the Cocoa Research Institute of Nigeria (CRIN) and the Institute Francaise du Cacao et du Cafe (IFCC).

The early development of the cacao industry in West Africa was entirely due to the initiative and entrepreneurship of the West African peasant farmers (Gibberd, 1951). In Nigeria, the Government has developed an interest in the cultivation of cacao since 1887 when cacao seedlings from the old Botanic Garden at Ebute Metta (Lagos) were sent up-country (Ibadan) for trial. This explains why cacao cultivation gained its first and earliest impetus around Ibadan, Oyo State of Nigeria. In Ghana, Teteh Quashi planted the cacao seeds he introduced in a village west of Accra. The cultivation of cacao developed around the village and expanded through Koforidua, Tafo to the Ashanti region which is today the main cacao producing area of Ghana. In the Ivory Coast, the first cacao plantings were at Bingerville, just outside Abidjan. The bulk of cacao production in the Ivory Coast currently comes from the forest region of Divo. Sierra Leone, Togo, Republic du Benin and other West African countries embraced cacao cultivation with a lot of enthusiasm in the early part of the

Table 7 Development of African raw cacao bean export 1900–1970 ('000 tonnes)
Source: *Cocoa Statistics*. Gill & Duffus Group, London, 1975

Countries	1900	1910	1920	1930	1940	1950	1960	1970
Cameroun	—	3	3	10	24	47	59	72
Gabon and Congo	—	1	1	1	2	4	4	6
Ghana	1	23	127	194	228	272	308	367
Guinea	1	2	7	6	16	15	33	19
Ivory Coast	—	—	1	22	46	62	63	143
Nigeria	—	3	17	53	91	102	160	196
San Thome and Principe	14	37	20	9	7	8	10	10
Sierra Leone	—	—	—	—	1	2	3	5
Togo Republic	—	—	3	6	5	4	9	31
Zaire Republic	—	—	—	—	—	—	5	4
Other Africa	—	—	1	1	2	1	2	12
Total (Africa)	16	69	179	305	422	517	656	865
Total (World)	95	216	370	482	614	747	896	1119

twentieth century. Not in all cases has the enthusiasm met with success (see Table 7). This is apparent from the evolution of the export of cocoa beans from Africa shown in Table 7 for the period from 1900 to 1970.

For a number of reasons including population increase, local need for food production, industrial development and shortage of fertile land, cacao production appears to have passed its peak in West Africa.

7.2 TAXONOMY

Cacao belongs to the genus *Theobroma* in the family of the Sterculiaceae. Over 20 species of *Theobroma* are recognized.

All cacao cultivated for the international market belongs to the single species *Theobroma Cacao* (L.). Other *Theobroma* species are locally exploited by the indigenous populations either for making refreshing drinks from the sweatens or for making a type of chocolate from the cotyledons.

There are three large and distinct groups within the species *T. cacao*. These are the Criollo, The Trinitario and the Forastero Amazonian.

7.2.1 The Criollo Group

The main characteristics of the Criollo cacao types are slender trees, green pods or pods coloured by anthocyanin pigments; warty, thin, soft pericarp, lignified mesocarp, beans plump and embedded in pulpy mucilage, white cotyledons. On fermentation and drying the cotyledon colour turns light brown.

The Criollo type is the most anciently cultivated cacao type. The beans possess excellent flavour. The trees are poor in vigour of growth and they are susceptible to disease. The commercial production of Criollo cacao types has remained restricted to around their centre of origin, namely Venezuela, Mexico, Nicaragua, Guatemala and Colombia. Experimental plantings were made elsewhere, but these have not yielded results that could encourage commercial planting of Criollo cacao types outside the new world.

7.2.2 The Trinitario Group

The Trinitario group contains hybird populations of mainly Forastero Amazonian and Criollo. As a hybrid population which has undergone generations of segregation, the Trinitario group is very variable and polymorphic. The high level of heterogeneity of its genotype has conferred a high degree of environmental plastic adaptability to this group.

Generally, the trees are vigorous with a variable disease reaction. Pods are green or pigmented. The beans vary in colour from very light to very dark purple. Although the initial popularization of the Trinitario was in Central America and Trinidad, it is now commercially cultivated along with other types in most cacao growing countries of the world.

7.2.3 The Forastero Amazonian Group

This group originated in the Upper Amazon Basin, and it contains most of the cacao commercially grown in Brazil, West Africa, Central America and the Caribbean Islands. Selections were made in Trinidad from material collected by Pound in 1938 and 1943 when he was searching for trees with resistance against *Marasmius perniciosus*. Posnette A. F. (1943) in a search for resistance to the West Africa cacao viruses, introduced these Upper Amazon varieties to Tafo, Ghana.

The Forastero group is characterized by green pods, absence of anthocyanin pigmentation, thick pericarp, strongly lignified mesocarp, plump but slightly flattened beans, deep purple cotyledons when fresh.

The Forastero cacaos have done very well in West Africa mainly on account of their superior vigour of growth, precocity, mild to strong tolerance to West African virus strains and high bean yields.

7.3 BOTANY

Theobroma cacao is cauliflorous and semi-deciduous. The tree is low, reaching an average height of 5 to 10 m. The main trunk is short, branching (jorquetting) in whorls of five branches. Branches are dimorphic:

1. Verticals or chupons growing from the trunk have leaves arranged in $\frac{5}{8}$ phyllotaxy.
2. Lateral branches (fans) with $\frac{1}{2}$ phyllotaxy. Petioles with two joined pulvini, one at the base and the other at the point of insertion of the leaf. Stipules 2, deciduous. Lamina elliptical – oblong or obovate–oblong, simple, 10 to 45 cm long; generally smooth, sometimes hairy, rounded and obtuse at the base, pointed apex. Dichasial inflorescence, primary peduncle very short, often thick and lignified. Flower peduncle 1 to 4 cm long. Sepals 5, triangular, whitish or reddish in colour. Petals 5, basally joined into a cup-like structure, whitish yellow with two dark purple bands adaxially; ligules spathulate, yellowish. Stamens 5, fertile, alternating with 5 staminodes, the two whorls uniting to form a tube. Anthers 2, fused stamens. Ovary superior with a single style terminating in 5 sticky stigmatic surfaces. Fruit variable in shape, ovoid, oblong. Sometimes pointed and constricted at the base or almost spherical, 10-furrowed out of which 5 are prominent. Axial placentation, seeds embedded in mucilage, flat or round with white or purple cotyledons (see Figure 21).

T. cacao is naturally outbreeding. Various insects have been associated with pollination in cacao. The main pollinators are thrips, midges, ants and aphids. As an outbreeder, cacao possesses a complicated system of self-incompatibility to ensure cross-breeding. Self-incompatibility in *T. cacao* is genetically controlled in both the gametophytic and the sporophytic systems.

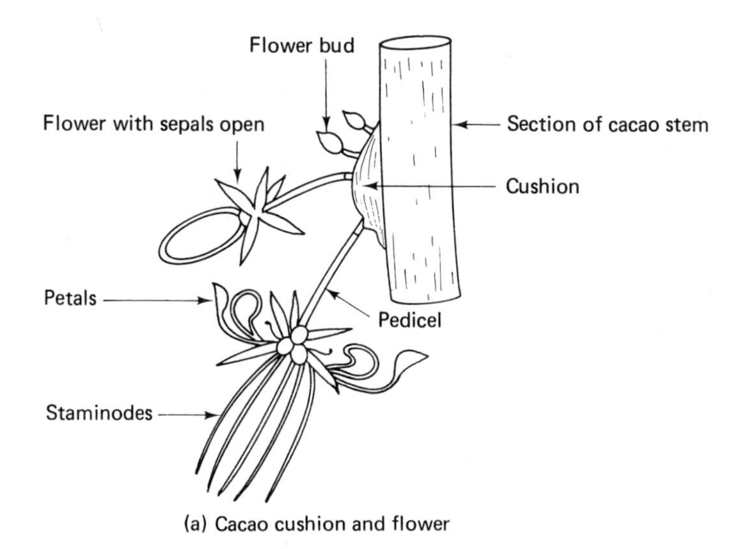

(a) Cacao cushion and flower

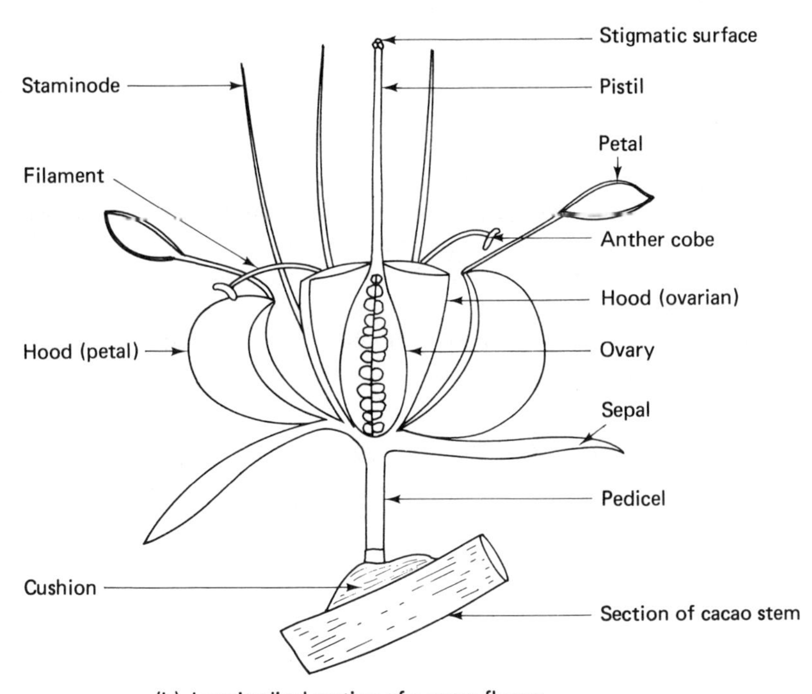

(b) Longitudinal section of a cacao flower

Figure 21. Cacao cushion and flower

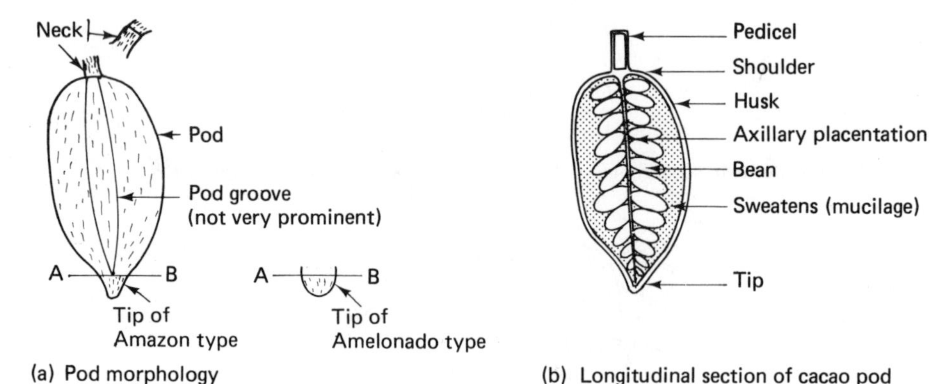

(a) Pod morphology

(b) Longitudinal section of cacao pod

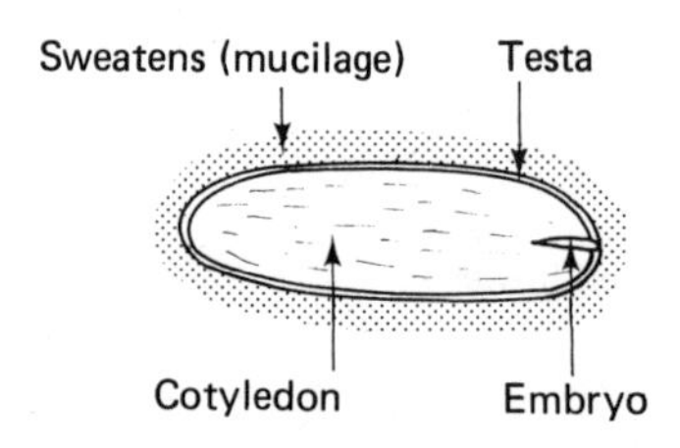

(c) Longitudinal section of a cacao bean

Figure 22. Morphology of cacao pod and bean

After successful pollination, fertilization takes place within 36 hours; the sepals, petals and staminodes drop away; the stamens and the pistil wither. The young pod, known as cherelle, commences development by longitudinal elongation, followed by increase in girth. The period between fertilization and pod maturation varies from 150 to 180 days depending on the variety. In all varieties, the pod colour turns light yellow when ripe. Pods are ready for harvesting at this stage (see Figure 22).

7.4 AGRONOMY OF CACAO

7.4.1 Sitting the Plantation

7.4.1.1 Climatic factors

Cacao is a low-altitude crop. It can grow from sea-level up to an altitude of 700 m. In West Africa, the best crops have been produced within altitudes of 100 to 300 m.

Cacao can be grown within a wide range of rainfall from 1000 to 3000 mm p.a. or more. When irrigation is available, and provided that the occurrence of dry winds is limited, the crop can be grown completely without rain. Cacao is very sensitive to moisture stress. It is also very sensitive to excess water in the soil.

Rainfall also has a profound effect on relative humidity, temperature and evaporation. The minimum loss of soil water through evapotranspiration as estimated from measurements taken over a large number of cacao plantations in many cacao growing countries is of the order of 100 to 125 mm per month (Hardy, 1960). This means that the minimum requirement to replace water loss through evapotranspiration is of the order of 125 mm monthly: unless this is obtainable soil water will be depleted especially during dry months of the year. Therefore, in selecting a site for planting cacao, it is desirable to ensure that the site enjoys rainfall averaging 150 mm per month for at least nine months of the year, i.e. a minimum rainfall of 1350 mm per annum distributed as evenly as possible over nine months of the year.

The optimal average temperature throughout the year for cacao cultivation is around 25.5 °C with a daily range of not more than 10 °C. It has been suggested that daily temperature fluctuations of this order are necessary to initiate bud bursting. However, bud bursting and flushing become excessive when daily temperature ranges are in excess of 10 °C. Low temperatures adversely affect cacao trees. Hardy (1960) and Erneholm (1948) suggested that 15 °C should be the lowest daily temperature for cacao. At average temperatures below 25.5 °C flower formation in cacao is inhibited and trunk growth appreciably reduced.

Ideally, seasonal variation in temperature should be small and the diurnal temperature range almost constant throughout the year. Away from the equator, diurnal temperature ranges increase, and conditions become progressively more unsuitable for growing cacao. The bulk of the world cacao is produced within 10° of latitude from the equator, at low altitudes. Some cacao is cultivated in Mexico (20 °N) and Brazil (30 °S). Cacao cannot tolerate frost.

However, during the dry season the occurrence of harmattan winds bringing air of very low relative humidity, may cause serious defoliation of cacao trees. Otherwise, high humidity and an overcast sky may serve to ameliorate adverse effects of long periods of very low rainfall. This tends to explain the success of cacao in some parts of the 'middle belt' of West Africa. However, there is a limit to this amelioration effect as casual observation has shown that cacao cultivation is uneconomic in the low humidity–low rainfall northern areas of the cacao-growing countries of West Africa. Where soil moisture is adequate, high air humidity is undesirable for cacao, as this favours the spread of one of the most important diseases of cacao – pod rot, caused by *Phytophthora palmiwora* (Butler).

Cacao is very susceptible to damage by strong winds. Cacao roots are superficial, thus offering little resistance to strong winds, which also cause considerable damage to the fragile branches, the young leaves, and the flowers of cacao. Dry winds, such as the harmattan, also cause dehydration of floral organs, rendering them incapable of opening, thus resulting in failure of pollination. In most parts of West Africa, the dry harmattan winds are prevalent from December to February.

Table 8 Analytical data of cacao soils derived from metamorphic rock in Western Nigeria (Wessel, 1966)

Depth (cm)	Soil colour (wet sample)	Coarse sand	Fine sand	Silt	Clay	pH	m.e./100 g fine earth K	Ca	Mg	Cation exchange capacity	%C	%N	'Available' P in p.p.m.
Soil derived from fine grained biotite gneiss; profile under cacao, near Ile-Ife, Nigeria (mean annual rainfall 1270 mm)													
0–5	10 YR $\frac{3}{4}$	31	47	6	15	6.5	0.20	9.8	1.90	13.3	3.13	0.21	14
5–18	7.5 YR $\frac{3}{4}$	26	44	5	25	6.1	0.18	5.0	1.00	8.6	1.46	0.10	14
18–48	5 YR 4/6	25	46	5	24	6.0	0.14	3.1	1.00	5.2	0.66	0.06	14
48–99	5 YR 4/8	20	47	7	26	6.1	0.15	2.7	0.90	5.0	—	—	16
99–135	5 YR 5/8	18	28	7	47	6.1	0.15	2.7	0.90	5.3	—	—	11
135–160	5 YR 5/8	28	20	6	46	6.0	0.15	2.9	0.80	5.2	—	—	17
Soil derived from medium grained granite rock; profile under cacao, near Owena, Nigeria (mean annual rainfall 1448 mm)													
0–8	10 YR 3/3	35	38	4	23	6.9	0.29	11.9	2.90	17.4	2.92	0.25	8
8–18	10 YR 4/4	30	46	7	17	5.9	0.16	3.1	1.30	5.7	0.80	0.08	13
18–33	7.5 YR 4/4	11	66	3	20	5.9	0.18	2.4	0.70	4.5	0.53	0.06	8
33–51	5 YR 4/4	10	49	7	34	5.8	0.26	2.5	0.80	7.0	—	—	9
51–81	5 YR 4/6	29	21	8	42	5.8	0.24	2.4	1.10	6.9	—	—	8
81–152	5 YR 4/8	21	23	8	48	5.7	0.19	2.3	1.70	6.6	—	—	9
Soil derived from coarse grained gneiss; profile under cacao, near Oke-Irun, Nigeria (mean annual rainfall 1346 mm)													
0–8	7.5 YR 3/2	44	31	8	17	6.1	0.50	5.7	1.50	10.3	1.94	0.15	10
8–15	7.5 YR 3/2	44	32	7	17	6.0	0.27	3.6	1.20	7.5	1.43	0.10	23
15–33	5 YR 3/4	32	33	6	29	5.5	0.41	2.6	1.80	8.6	1.00	0.07	10
33–89	5 YR 4/4	29	24	8	39	5.2	0.49	1.9	1.70	7.4	—	—	9
89–127	5 YR 4/6	26	26	8	40	5.3	0.28	1.7	0.80	6.4	—	—	17
127–152	5 YR 4/6	32	26	8	34	5.1	0.22	1.8	0.90	5.1	—	—	12

7.4.1.2 Soils

Cacao is a tap-rooted plant and requires deep well-drained soils, free from iron concretions, high in nutrient content and a topsoil rich in organic matter. Cacao soils should have adequate clay content.

The general physical characteristics developing on the upper slope profile of soil catenas tend to form the best soil type for cacao. Upper slope profiles usually have a well-defined horizon at a depth of 250 to 1250 mm. Typical profiles of selected cacao soils in Nigeria are listed in Table 8.

7.4.2 Raising Cacao in the Nursery

7.4.2.1 Raising of seedlings

Cacao seeds readily germinate when sown and do not pass through a dormancy period. They lose viability on extraction from the pod within five to seven days, unless specially treated.

Cacao seeds are, therefore, best stored in the pods, where they remain viable for up to four weeks after harvesting. If, however, it is necessary to extract the seeds from the pods for storage purposes, the extracted seeds should be mixed with moist fine sand, moist sawdust or moist ground charcoal. The mixture should then be stored in a cool place. Wooden boxes are preferable to closed jars or tins because they permit some aeration for the respiration of the living seeds. Under such conditions, extracted cacao seeds can be stored for two to three weeks, but invariably some of the seeds already germinate within this period.

In nursery beds single seeds are planted per hole, which are usually 15 to 20 mm deep, and 15 to 16 cm apart. After planting, the seeds should be watered lightly every day. The seeds will germinate in 7 to 10 days. After germination, the amount of water applied should be increased, depending on the quantity of rain prevailing. It is advisable to confine watering to the mornings and evenings.

Cacao seeds can also be sown in polybags. These should measure 20 × 12 cm and have drainage holes. They are filled with fertile sifted topsoil. The filled bags are watered heavily the night before the seeds are sow. Seeds are sown one per bag. Light watering is done after sowing, daily until germination (7 to 10 days). Adequate water should be provided after germination.

Once germinated, the seedlings should be provided with shade. A regular examination for diseases and pests should be arranged. Seedlings which show symptoms of virus and/or bacterial infection must be uprooted and burnt in the incinerator. Fungal and insect attacks are controlled with chemicals. There is no need to apply fertilizer to cacao seedlings in the nursery.

About a week before transplanting the seedlings in to the field, seedlings which are raised in nursery beds should be partially dug *in situ* as shown in Figure 23.

This preliminary partial digging *in situ* stimulates the initiation of new roots prior to transplanting into the field. This treatment necessitates extreme care

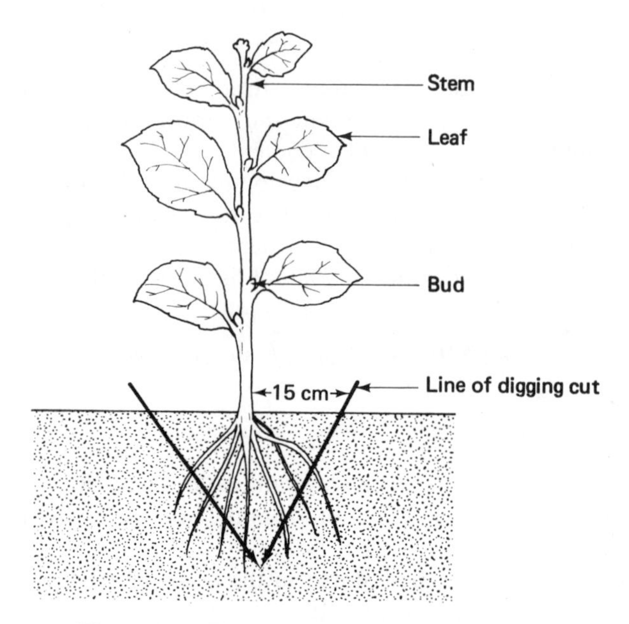

Figure 23. Partially dug seedling in nursery

during final digging to ensure that fragile young roots are not damaged.

If the planting site is near, each seedling should be lifted and planted with a ball of earth. Where long distance transportation of seedlings in unavoidable, seedlings will have to be transplanted 'naked root' but treated with a clay slurry. Seedlings which are raised in polybags do not require this treatment.

All seedlings should be sprayed against diseases and pests prior to transplanting into the field.

Some general recommendations for seedlings in nurseries are as follows:

1. Nurseries should be sited on clear-felled land near a permanent water supply; the seedlings should be provided with artificial shade of palm fronds, using 8 to 10 palm fronds per 3 m length of nursery bed.
2. The seed should be sown in pots containing forest topsoil; actual sowing should not be done at the hottest time of the day. Sowing should be done in December to February to allow the seedlings a period of four months of growth in the nursery before transplanting.
3. The pots should be watered thoroughly when the seed is sown, but thereafter, lightly every second day for the remainder of the dry season, except during harmattan periods when watering every day may be necessary.
4. The palm frond shade will dry out naturally, but should be removed in two or three stages, all the shade being removed about a week before transplanting.
5. Seedlings should be no more than four to five months old when

transplanted. Care must be taken not to damage the seedlings on the way from the nursery to the field, and planting of seedlings in flush should be avoided.

7.4.2.2 Vegetative propagation of Cacao

Cacao can easily be vegetatively propagated by leaf bud cuttings, multiple bud cuttings, marcotting, budding, grafting and layering. Rooting of cacao cuttings was initially carried out in concrete rooting bins. Nowadays, cacao cuttings are rooted by the polythene sheet method.

The floor of the rooting site is concreted; filled polythene bags are arranged on the floor in four rows of 10 to 12 pots. A dense shade is erected over the rooting site. The general routine of the operation follows the standard procedure except that the following points should be noted:

1. Fan cuttings should be taken for rooting, because these continue the sympodial adult pattern of growth, when rooted.
2. Cuttings should be taken early in the morning of the setting.
3. The stem base of the cutting should not be thicker than the stem of a lead pencil.
4. Watering of the cuttings should be done only to saturate the air inside the polythene sheet with moisture. A fine jet sprayer should be used for this purpose.
5. After rooting, the cuttings should be hardened off and subsequently kept in the nursery until the time of transplanting.

The propagation of cacao by cuttings has received considerable research attention in West Africa. Experience has shown that Upper Amazon and Trinitario type clones are the best rooters. It is less easy to obtain cuttings from the Amelonadoes and these are difficult to root. Local non-Amelonado types are intermediate in behaviour.

Some important precautions for a successful nursery for the estabilshment of rooted cuttings of cacao are as follows:

1. The shade provided should be uniform and made up of 12 to 15 palm fronds per 3 m length of nursery bed.
2. As cuttings are set in sawdust, this should be amply available. It should be allowed to have weathered for at least three months before use.
3. Cuttings are to be set early in the day.
4. Water only with a view to saturating the air inside the polythene cover with moisture. Apply water through a knapsack sprayer or a special watering can with fine nozzles.
5. The successfully hardened cuttings should be moved to a nursery bed and the plants should henceforth be treated like newly sown seedlings except that the shade should not be removed entirely until one week before planting in the field.

6. Plant the cuttings in a field with well-established temporary shade at the beginning of May, when the early rains become steady.

7.4.3 Transplanting

Cacao seedlings should be transplanted into a well-prepared field after they have grown in the nursery for about four to five months and when the rains are steady and regular. Older and bigger seedlings may look very impressive in the nursery; they are, however, likely to suffer a greater setback when transplanted into the field. It has been established that it is more economical to plant cacao either into selectively thinned forest or along traces cut through regrown forest. The generally accepted spacing for cacao in West Africa is 3×3 m. Trials are in progress to determine the optimal spacings for the newly bred and more vigorous Amazon hybirds.

When seedlings have been raised in nursery beds, they should be partially dug a week before the date of transplanting. A day before transplanting, the seedlings are dug up with balls of earth if the nursery site is close to the field, but with naked roots if seedlings have to be transported over some distance. The naked roots are treated with a clay slurry before being transported to the planting site. In both cases, seedlings should be planted immediately on arrival at the planting site. Under no circumstances should uprooted seedlings remain unplanted for more than 48 hours.

Seedlings raised in polybags do not require any special treatment before transplanting. They can also be transported over long distances without serious damage to the seedlings. Unlike seedlings with balls of earth or naked roots, polybag seedlings should be allowed a period of rest of not less than seven days after transportation over long distances (more than 5 to 6 km) before planting. This treatment has been shown on a number of occasions to improve considerably seedling survival after transplanting.

Great care should be taken to ensure that transplanting is properly carried out. Improper handling of seedlings during transplanting will invariably result in death of seedlings after transplanting. In most parts of West Africa, the best time for transplanting is late May or early June, the exact time depending on the location and the amount of rainfall in that particular season and area. In the wetter parts of West Africa, Cameroun and eastern parts of Nigeria, transplanting could be carried out as early as the month of April. Selection of seedlings for planting in the field is done in the nursery: small, badly-shaped and off-type seedlings must be discarded. It is a golden rule to increase the number of seedlings needed for planting in a field by as much as 5 per cent to make room for seedlings that may be damaged in transit and those that may be needed for early supplies. Seedlings with new flush growth should not be transplanted. The only method of avoiding transplanting many flushing seedlings is to observe the flushing patterns of the nursery seedlings and to adjust the transplanting time to coincide with the time that only a few plants are in flush.

Transplanting is a two-man operation. One man holds the seedlings in position while the second man removes the polybag or arranges the seedling roots in position, fills in the soil and consolidates. When seedlings are raised in cane baskets, they are planted with the baskets intact.

Cacao seedlings have indeterminate monopodial and rhythmic flushes of growth for the first 18 to 24 months after transplanting into the field. Subsequently, the seedling jorquettes, i.e. the terminal bud breaks into five lateral branches which develop to form the first storey of the plant. Growth of the jorquettes is sympodial. As the jorquettes develop, a vigorously growing chupon arises from the base of the jorquette, assumes a monopodial growth, and after growing for a season or a little more, breaks the terminal bud into five to give rise to the second jorquette which forms the second storey of the plant. The third storey which is less prominent arises in a similar manner after which the whole plant continues to grow sympodially. Precocious seedlings commence to flower around the same time that they form a jorquette.

7.4.4 Maintenance of the Cacao Farm

The maintenance of a cacao farm is an all-year programme embracing various operations at different periods of the year. The operations vary with factors such as climate, fertility of the soil, age and vigor of the cacao trees and method of establishment of the cacao farm. If the maintenance is correctly and adequately carried out, diseases and weeds will be less of a problem and their control can be achieved far more efficiently and effectively.

7.4.4.1 Weeding

Weeding is a general problem involving the removal of annual plants and shrubs mainly in the first three to four years after establishing cacao in the field when the canopy is not yet closed. Unless the weeds are properly controlled, they compete for water and nutrients and cause poor growth of the cacao trees. Frequency of weeding will depend to a large extent on the amount of overhead shade and rainfall. Clean-weeding, which is ideal, may be possible on smallholdings of up to 2 ha, but is certainly not feasible on plantations. In the latter case, row weeding thrice a year, supplemented by at least four to six slashings of the avenues per year is considered adequate. Weeding may reduce the incidence of black pod disease as it allows for a better circulation of air, thus helping to reduce the humidity within the cocoa farm. It also helps in the control of insects, especially leaf-eating caterpillars. Results obtained from trials on weedkillers (herbicides) are encouraging and with the increasing cost of labour the use of herbicides has every chance of becoming a standard practice especially on large plantations. With the current level of costs, available weedkillers have considerable advantages and are economic. The main disadvantages in the use of weedkillers are persistence in the soil and high mammalian toxicity. There are, however, no proven records of instances where these have limited the use of weedkillers. The following herbicides have been recommended after trial in different parts of West Africa:

1. Aminotriazole
2. Simazine
3. Aminotriazole + Simazine
4. Paraquat.

On a cost basis, paraquat appears to be the best, but it must be applied before the weeds blossom.

7.4.4.2 Mulching

Young cacao trees should be mulched before the onset of the first dry season to assist in conserving moisture in the soil. Mulching helps to reduce losses of seedlings that occur during the first and perhaps the second year of planting in the field. Mulching also helps in the maintenance of soil fertility by acting as an anti-erosion measure and by conserving organic matter; it reduces loss of soil water by decreasing evaporation. With mulching added to other essential cultural operations, plant growth is superior and this results in the earlier formation of a closed canopy; this suppresses weeds and thereby reduces costs of maintaining the plantation.

Each plant should be given a 15 cm deep layer of mulch towards the end of the rains, but a small area around the base of each seedling should be left clear to reduce termite attack on the stem. Grass or polythene sheets are good for mulching, but grass is preferred because of the high cost of polythene. Polythene sheet, however, helps to conserve more moisture. Avoid woody trash *close* to cocoa plants because of the risk of attack by termites.

7.4.4.3 Shade

Temporary shade (nurse shade) is beneficial during the early years before the cocoa canopy closes. The ideal nurse shade for cacao should be easy to establish and provide within a short time a good overhead shade throughout the dry season. It should compete as little as possible for moisture and soil nutrients with the cocoa roots and be easily removed when no longer needed; however, its removal should not damage the cacao canopy. It is important that the shade plants are not an alternative host species to insect pests or diseases of cocoa.

Plantain appears to meet most of the requirements and, for the moment, this is recommended. Banana should be avoided as it competes much too severely with cocoa for moisture during the dry season. Nurse shade should be thinned after a couple of years but its complete removal will depend on when the cocoa has formed a closed canopy. Both thinning and removal of nurse shade should be done at the end of the dry season. Felling of timber trees in a mature cocoa plantation should be discouraged. An alternative and possibly better method of shade reduction is the use if tree poisons. However, branches falling from the dying shade tree may cause considerable damage to the cocoa.

7.4.4.4 Pruning

Pruning operations are a generally accepted practice and are carried out essentially to remove unwanted growth and thus maintain regularly shaped trees carrying a well-balanced, firmly closed umbrella-shaped canopy about seven to eight feet (2.5 m) above the ground level. Pruning is sometimes done with a view to rejuvenating old trees. Rash and unskilful heavy pruning may lead to unnecessary injuries to trees with consequent reduction in yield. Light pruning is recommended although the frequency of pruning depends on the growth of the trees. Unwanted vegetative parts should be removed as close as possible to the stem on which they have grown, preferably at an early stage to save the trees from wasting their energy on unwanted growth.

An already mature cocoa farm which has never been pruned can be transformed into a good farm with well-shaped trees with a good canopy by a general cleaning up process which aims at the removal of low hanging, broken and dead branches and the regeneration of fallen or damaged trees. If the trees are still vigorous and the pruning goes on with other maintenance practices, a good cocoa farm with a closed canopy can be achieved in about three years. At this stage, canopy maintenance and regular chupon removal should be carried out.

Canopy maintenance and light pruning may be done at any time of the year, while heavy 'formation' pruning (to give a balanced symmetrical tree) should be done in February to March or May to June.

A lightweight sharp cutlass is suitable for most pruning operations. A sharp knife should be used for removal of chupons and young flushes on stems and major branches, while a heavy bowsaw is recommended for removing big branches and stems. Pruned surfaces bigger than 20 mm in diameter should be painted with ordinary paint or a tar sealing compound.

7.4.4.5 Regeneration

During storms heavy trees may be blown over or trees may be damaged by breaking of the main branches. Dormant buds soon become active and grow out rapidly. Three regularly spaced vigorous chupons at the base of the trunk should be retained but probably only one, or at most two, should eventually be allowed. The others should be cut off as soon as one chupon jorquettes.

When the fan branches are about 45 cm long, the top should be cut in order to encourage secondary branching. Thus, a well-balanced canopy will develop. Earth is heaped up around the base of the new shoot to encourage growth of an independent root system. Chupons growing closest to the ground are preferable to those growing higher as it is easier to earth them up.

The most difficult part of the operation is to fit the young plant into the overall canopy. In most cases, the new tree will flower and bear its first pods about two years after the new buds start to grow. The surrounding trees should be prevented from choking up the new tree by regular removal of their growth into the gap.

Regeneration is an important operation designed to maintain continuity of yields whatever the age of the cocoa farm.

7.4.4.6 Replacement of dead seedlings

Supplying of vacancies is usually done during the first two rainy seasons after field planting of the new farm. This will ensure a measure of uniformity among the trees. Supplying operations should be done as soon as the rains become regular, usually from late April to early June. The essential requirement is adequate rainfall during and after the operation. Supplying operations carried out late in the rainy season often allow the seedlings a short spell of rainfall followed by a period of dry weather and rather heavy casualties may occur. It is relevant to mention that the earlier the vacancies can be filled, the better are the chances of these seedlings catching up to give a uniform cocoa plantation.

7.4.4.7 Fertilizer application

Results of fertilizer trials with cacao in West Africa have shown that fertilizer application to cacao can be economic. Fertilizer is becoming increasingly important in cacao cultivation in West Africa because virgin forest soils for cacao growing have been exhausted. New cacao farms have to be established in depleted secondary bush soils and in a few cases such newly established farms have to be intercropped with food crops. The most important fertilizers needed in cacao are nitrogen (N), phosphorus (P), potassium (K) and boron (B).

Field experience and preliminary pot experiments show that in most of the soils under fallow or forest cover in West Africa soil nutrient level is not a limiting factor to the growth of cacao provided that the structure of the soil the water retention are adequate. This is where a soils map of the cacao growing belt is very useful in selecting a new site for cacao. A cacao farm sited on any of the soils classified as good for cacao growing and which has not been previously cultivated, will not respond to fertilizer application. Most of the cacao farms now requiring replanting or rehabilitation either have been on marginal soils, or have been carrying crops for many years during which the nutrient status must have declined. In such rehabilitation work the use of fertilizer is likely to be of advantage.

It pays to apply fertilizers to mature cacao trees especially after they have been in bearing for a number of years. Nutrient deficiency in mature cacao trees can be easily detected and the need of cacao for fertilizers can be predicted through appropriate diagnostic methods and use of leaf analysis data.

The economics of fertilizer application to cacao in West Africa was discussed by Wessel (1966cm) who gave a likely average of 30 per cent yield increase from the application of combined nitrogen and phosphorus fertilizers. Although there is no doubt that fertilizer use is of economic advantage, the actual return in terms of cash value is still difficult to evaluate. Apart from the variation of the price of cacao from year to year, and the insufficient information about the handling charges of the fertilizer, the nature of the response to fertilizers depends on many factors such as the physical condition of the soil, its history, the age of the plant and the general management.

The increases in the yield so far obtained due to nitrogen and phosphorus fertilizer application vary from 3 to 54 per cent; this is a reflection of the variability in the age of the cacao and in the chemical and physical conditions of the soil. About 35 to 40 per cent of the present farmers' cacao holdings in this area are sited on soils physically unsuitable for cacao growing. The application of fertilizer in such cases will be of no economic advantage without improving the physical conditions of the soil.

Cacao farms should be weeded or the weeds well cutlassed before fertilizers are applied. With young cacao trees with undeveloped canopy, the fertilizer mixture is applied in a circular band around the tree on the surface of the soil at a distance of 10 cm from the plant. In the case of mature cacao, however, the first effort is to ensure that the deep cacao leaf litter does not prevent the fertilizer from reaching the soil in time. Where the leaf litter is thick and compacted, it pays to loosen the compaction prior to fertilizer application. Fertilizers are generally broadcast under mature cacao trees. Direct contact of fertilizer with the trunk of the cacao trees should be avoided. The essential rules of fertilizer application to cacao are as follows:

1. Always cutlass the farm before application of fertilizers.
2. Protect the trees against capsids.
3. Protect pods against black pod disease.

7.5 CACAO DISEASES AND THEIR CONTROL

When a high standard of sanitation is maintained, incidence of diseases and pests is greatly reduced in tree crop plantations. This is very true of cacao farms. Nevertheless, several diseases may cause considerable harm.

7.5.1 Swollen Shoot Disease of Cacao

Swollen shoot disease of cacao is a virus disease. It was first noticed in Nigeria in 1944 and similar diseases have been found in cacao growing in Ghana, Ivory Coast and Sierra Leone. About 20 strains of the virus, which are recognized by the symptoms and the severity of their effect on yield, are known in West Africa. The disease has been one of the most important factors limiting cacao production. Over $1\frac{1}{2}$ million infected trees were destroyed between 1946 and 1956 in an unsuccessful attempt to eradicate the disease. Cutting out is still used to keep the disease under control in most parts of West Africa but not in the areas of mass infection (AMI), where control measures and inspection have been abandoned. The AMI make up one-sixth of the cacao growing area of Nigeria, one-fourth of Ghana and none in the Ivory Coast. In these areas the virus in combination with mirids and environmental factors takes its greatest toll.

7.5.1.1 Symptoms

The first sign may not appear until six months or more after a cacao tree is infected. Symptoms are seen on tissue produced after infection, most frequently on the rapidly growing shoots, not only on the leaves and stems, but also on the pods and roots. The first symptoms appear on young flush leaves as a network of red vein banding which soon develops into vein clearing or chlorosis. Red vein banding is a condition in which the veins of the young leaves and the area along them are red instead of green. Vein clearing on the other hand means that the veins are yellow and one can almost see through them. Chlorosis is a general yellowing effect of the leaves between the veins.

All stem and root symptoms appear as swellings. With some virus strains, however, swelling may also develop on fan branches or chupons. Infected pods appear rounded and small or show some mottling.

The first symptoms are very conspicuous and this phase is followed by a chronic or 'recovery' phase, with few symptoms. Symptoms are rather difficult to find in the dry season since leaves with symptoms usually drop and shoots with swellings are often destroyed by drought and die-back.

7.5.1.2 Effects on growth and yield

The growth of cacao trees infected with swollen shoot disease is generally reduced. The disease reduces stem diameter, seedling height, number of leaves, root length and number of roots, but has no effect on stem/root ratio. If mirids are also attacking the trees the canopy of a cacao farm may be destroyed and eventually weeds invade the farm and compete with the weakened cacao trees which ultimately die. It is thought that if mirids are controlled, swollen shoot disease is not fatal and the reduction in yield would be from 5 to 20 per cent. Nevertheless, virus infection remains a serious threat to the cacao industry in West Africa.

7.5.1.3 Spread

The virus which causes the disease is spread by small white insects called mealybugs, and cannot be carried by soil or in cacao beans (seeds). Having fed on an infected tree the mealybug can infect any healthy tree it feeds on during the next 36 hours. The mobile immature mealybugs or 'crawlers' and particularly the first instar nymphs are mainly responsible for carrying the disease from one cacao plant to another. The adult mealybugs rarely move and are considered of little importance in the transmission of the virus. Their importance lies in their ability to produce, within a short period of time, very many fast-moving crawlers which spread the disease. Three main types of spread are recognized. The first is by mealybugs walking from the canopy of an infected cacao tree to that of a healthy neighbouring cacao tree and feeding there. This type of movement, which is termed, '**radial**' or '**continuous**' spread, is commonest at the beginning and end of the wet season. Unaided passage from tree to tree along the ground appears to be unimportant because mealybugs move slowly and have many enemies on the ground, such as lizards and spiders. The second type of spread is by infective mealybugs carried by the wind from infected to healthy trees; this is known as '**discontinuous**' or '**jump**' spread, and is responsible for most spread over long distances. Dispersal by wind appears to be commonest in the dry season, and most new outbreaks of virus probably start at this time.

But spread could also take place when mealybugs are carried by ants from infected to healthy trees. Ants play an important part in virus transmission because in addition to transporting mealybugs, they protect the mealybug colonies by covers, or 'tents' of soil and plant debris; these act as shields from attack by parasites and predators. The ants also remove moulds (fungi) which might kill the mealybugs. Without these attendant ants, there would be fewer mealybugs and virus spread would be slower.

It is also possible that mealybugs feeding on pods taken from infected trees can transmit the virus. In this way virus may be carried to a healthy part of a farm during harvesting or from the wilting pods and decomposing husks discarded after the beans have been removed.

However, not only cacao trees harbour the disease. Wild kola, silk cotton tree and several other trees contain the swollen shoot virus which can be transferred to cacao by mealybugs.

The rate of spread of this virus disease varies from site to site, probably due to the following factors:

1. The size of the outbreak.
2. The size, composition, distribution, species and behaviour of the mealybug population.
3. The length of time the insects spend on the infected plant.
4. The spacing, age, variety, tolerance to infection and vigour of the cacao trees.
5. The virus strains.

7.5.1.4 Types of virus

Each strain of virus causing swollen shoot disease in West Africa is named after the place where it was first found. The most damaging strains include those from Egbeda (or Ikire), Offa-Igbo, Busogboro and Ilaro in Nigeria and the Jubean strains in Ghana. Some of the strains in Ghana can kill a cacao tree in less than a year, but the most severe strains in Nigeria, namely Egbeda and

Offa-Igbo, kill much more slowly unless the cacao trees are also attacked by mirids and the associated die-back fungus, *Calonectria rigidiuscula*.

Cacao trees are attacked by two other virus diseases, namely cacao necrosis virus (CNV) and cacao mottle leaf virus (CMLV), but the importance of these two viruses has not been critically assessed.

7.5.1.5 Control

There is no cure for virus diseases of cacao. The only possible method of control is to cut out (remove) all infected trees. Inspection for infected trees and their destruction are usually carried out by the cacao survey teams. Since some recently infected trees may not show symptoms, outbreaks are treated according to the following recommendations: when 1 to 5 cacao trees are diseased, cut out all trees up to 4.5 m from these; when 6 to 49 cacao trees in a group are diseased, cut out all trees up to 9 m from these; when more than 50 cacao trees in a group are diseased, cut out all trees up to 14 m from these. This treatment should remove all recently infected trees in contact but may not remove subsidiary outbreaks arising from 'jump' spread, and one or more re-inspections of the area are necessary to detect these. Inspections are carried out at six-monthly intervals.

It is difficult to kill the mealybugs with insecticides because these cannot penetrate both the protective tents built by the attendant ants and the protective wax covering the insect's body. It is, however, possible to kill the attendant ants with insecticides but this always results in heavy damage to the cacao tree by stem borers, pod miners and other insect pests which are normally unimportant.

In the area of mass infection (AMI) field experiments have shown that whenever less than 30 per cent of the trees are infected, economic cacao yields can still be obtained if the farm is well maintained, i.e. if all weeds are removed and mirids (capsids) are controlled. In such circumstances, virus infection would be just another debilitating factor of less importance than the other influencing growth and yield of cacao trees. Improved varieties of cacao are being produced and these should be vigorous and continue to give high yields even when infected by the swollen shoot virus. Tolerance of the virus or even resistance in cacao trees would be the best answer to the debilitating virus diseases.

All cacao pods, budwood and other cacao materials imported into West Africa should be certified virus free in the country of origin and should undergo strict quarantine on arriving in West Africa to prevent the introduction of any other dangerous viruses or strains of existing viruses which might ruin the cacao industry. It is, of course, also important when selecting materials within West Africa for distribution, to ensure that such materials are virus free.

7.5.2 Black Pod Disease

Black pod disease is the most serious disease of cacao in West Africa, especially in Nigeria and in Cameroun. It is caused by a fungus called *Phytophthora palmivora* and occurs only during the wet season. The disease is worse in areas of heavier rainfall (Opeke and Gorenz, 1974). The major damage from the disease is the rotting of both small and large pods. Chupons, seedlings (in nursery beds) and leaves of trees are attacked and killed under specially severe disease conditions following long periods of cool, rainy weather.

The fungus lives through the dry season in the soil, especially at the base of the tree and in the interior of heaps of old pod husks. The first infected pods of the new season are usually found at the base of the tree touching the soil or within rain splash distance of it.

7.5.2.1 Symptoms

The first sign of infection of a pod is a small brown spot, with an irregular, somewhat water-soaked margin. This spot gradually increases in size and after two to three days, a thin whitish downy mould appears in its centre. The brown spot area enlarges and so does the area covered by the white mould until the whole pod and later also the beans have rotted. This may take up to 18 days for mature normal-sized pods and about four days for very young pods (cherelles). At this stage, other moulds establish themselves on the already dead pod tissue, covering it with a thick cotton-like blanket. In the meantime, the pod has turned black and finally dries out remaining on the tree as a lightweight, shrivelled and rather hard fruit, the interior of which is gradually eaten by insects. The disease derives the popular name of black pod from this stage of pod infection although such a black, shrivelled pod is not at all unique to black pod disease. Pods dying from other causes usually take on the same blackish appearance. The really characteristic symptoms of this disease are the brown colour of the newly infected parts of the pod and the presence of the associated thin whitish downy mould.

The black pod fungus also causes symptoms of wilt on the flush leaves of cacao; the tips get brown and show a shrivelled appearance. This infection occurs frequently in areas of high rainfall when the trees are flushing abundantly. The infection stops as the leaves harden off. In many countries *Phytophthora palmivora* infects the bark of the trunk, sometimes so badly that the tree is killed. The symptoms appear on the lower trunk as a longitudinal sunken spot, sometimes 30 to 60 cm long. The importance of this infection is yet to be determined.

Finally it is known that *P. palmivora* is able to infect cacao roots and kill them. Under certain conditions one-month-old seedlings can be killed within five days of infection.

7.5.2.2 Spread

The white mould in the centre of the brown spot produces innumerable spores which can easily be carried to healthy pods – mainly by splashing or dripping raindrops – to cause new infections. That is why one very often finds newly infected pods just beneath an earlier infected one. Most spores, however, do not land on another pod but reach the cacao branch, trunk or soil. Spread through raindrops is mainly in a downward direction. Spread to pods in the top of the tree can be by squirrels, by insects or even by harvesting hooks or knives that have had contact with infected soil or diseased pods. When it is not raining, relatively few spores, if any, are scattered and no spores germinate unless they become humid or wet from dew or rain.

A period of sunny weather and less rainfall such as occurs in the short dry spell during the month of August in some parts of West Africa reduces spread of the disease. The disease does not spread any further when the rainy season has come to an end. It reappears slowly when the next rainy season starts.

7.5.2.3 Control

To be most effective, control should be aimed at preventing the appearance of the disease. This means that control measures should be started at the beginning of the rainy season before the disease appears. The fewer the number of trees on which black pod appears the more effective will be the spraying. The control measures for reducing pod rot are aimed at greatly reducing the number of spores available to cause new infections and/or preventing the spores which reach healthy pods from penetrating them.

7.5.2.3a Removal of infected pods

Old, dead pods on the tree from the past season are no longer infective and thus cannot cause more pod rot. Spores which may cause new infections are found mainly on the pods infected less than a month earlier and which have a large area covered with the white, downy mould. Hence, it is important to remove the first, newly infected pods that appear on the trees early in the season. Such pods should be picked off at the earliest detectable stage of the disease (evident as small brown areas) before the white mould appears and should be removed from the plantation. For this purpose, the trees should be inspected at least every week.

7.5.2.3b Weeding

Bushy farms tend to be more humid and thus favour easy germination of the spores which may cause new infections. Regular weeding and general maintenance is, therefore, essential for reducing black pod disease as it favours quicker drying of the surfaces of the pods.

7.5.2.3c Chemical control

For spraying to be successful, it should be started very early in the season in order to prevent the appearance of pod rot on the tree. It should be repeated every three weeks. Since the first infected pods appear at the base of the tree, all pods on the lower trunk should be given thorough and heavy spray application.

Many farmers have become discouraged with spraying because they started spraying only after the disease was well established in pods on the trees and as a result the spraying gave unsatisfactory control.

The purpose of using the fungicide is to kill or to prevent the germination of spores on contact. When the pods are sprayed and kept covered with such a chemical the amount of disease is considerably reduced. Rain, however, washes some of the chemical off and newly grown parts of the pods would not be covered with the protective chemical. Therefore, spraying must be repeated every three weeks, starting at the beginning of the rainy season, and continuing until the rains have ceased.

The crop protection chemicals currently recommended for use against this disease in West Africa are Perenox, Brestan, Carbide–Bordeaux mixture and Lime–Bordeaux mixture. They are all about equally effective. The Lime–Bordeaux mixture is the cheapest and is, therefore, preferred.

7.5.2.3d The use of Lime–Bordeaux mixture

To prepare 450 litres of spray, dissolve 4.5 kg of copper sulphate snow in 350 litres of water and 2 kg of hydrated lime in 100 litres of water. Mix the two solutions thoroughly by stirring, while pouring the copper solution into the lime. To prepare 45 litres of spray simply divide the above quantities by 10.

The most economical way for the farmer to spray these chemicals is by means of the ordinary pneumatic knapsack sprayer. Care must be taken to ensure that all pods are properly wetted on all sides.

If the spraying is done as recommended, the number of black pods will be reduced by half or more. For every 2 Naira spent on spraying (for materials and labour), one will obtain from the cocoa saved at least 4 Naira. On the best yielding farms the value of the cocoa saved can be up to four times the cost of spraying.

In spraying, it is the chemical that remains on the pod that protects it against pod rot. Because of the waxy surface and contrary to what is expected, more chemical remains on the pod if it is sprayed lightly and quickly so that the tiny drops of the chemical on the pod do not touch each other. If the drops touch each other, they come together, become rather large and run off, resulting in a waste of spray material.

Also, most sprayers have nozzles that spray too wide and have too high an output of the chemical. By using a nozzle with a small opening (0.7 mm) and lower pressure the amount of chemical used can be reduced by half or more. With such nozzles, fine screens are needed in the spray hose and they must be kept clean to prevent clogging.

7.5.3 Charcoal Rot

Charcoal rot is caused by *Botryodiplodia theobromae*. This fungus is a weak parasite and cannot infect vigorously growing pods. Only wounded, overripe or otherwise weakened pods are infected.

Initially the infected spot has a dark brown colour, but the tissue maintains its original shape. Soon, small pustules become visible but no aerial mycelium is formed. These pustules produce chains of spores, looking like thin greyish curly hairs. Within a few hours this grey colour turns deep black; hence its name. This disease is not of economic importance and no control measures have yet been worked out.

7.5.4 Fusarium Pod Rot

This is caused by a *Fusarium* species. This fungus also infects weakened pods only. The infected tissue, however, collapses after turning brown, so that the infection appears like a brown sunken spot. No aerial mycelia can be seen, nor small pustules with black spores; the fruiting bodies appear like white-pinkish, somewhat oblong cushions and contain innumerable spores. They are much the same as the fruiting bodies of *Calonectria rigidiuscula* on branches of trees suffering from 'die-back'.

7.5.5 Other Pod Rots

Two other pod rots have been recorded. Thielaviopsis pod rot is caused by *Ceratostomella paradoxa*. It is a minor disease for which no control measures are recommended. This is also true for a bacterial pod rot, caused by a *Bacterium* species.

7.5.6 Root Diseases

Root diseases occur on cacao when the causal fungi are deprived of their normal growing medium, e.g. (roots of forest or temporary shade trees) through the thinning of forest trees or by the removal of the temporary shade. The fungi, forced to look for other media on which to grow, find the nearby cacao root most suitable.

7.5.6.1 Collar crack disease

This disease is caused by *Armillaria mellea*. It can be recognized by the longitudinal cracks it produces in the lower trunk and roots of cacao. These cracks can be up to 75 cm long and are filled with fungal mycelia. In some cases the mycelia protrude outside the cracks. In addition, a continuous blanket of white mycelium can be seen between the bark and the wood of the trunk and roots, when the bark is peeled off carefully. In plantations, such trees are easily noticeable because of excessive defoliation.

7.5.6.2 Black root disease

This disease is caused by *Fomes noxius*. The field symptom of this disease is again excessive defoliation of leaves. The mycelium, however, does not grow under the bark but as separate strings on top of the roots and forms a continuous black-coloured crust on the surface of the roots and the lower trunk.

7.5.6.3 White root disease

White root disease is caused by *Fomes lignosus*. In this disease, white mycelia cover the root surfaces which lack cracks or crusts. Mushrooms are conspicuously visible around the base of cacao trunks near soil level. These mushrooms are brown on the upper side and creamy yellow on the under side. Very often they can also be found on decaying stumps surronding the diseased trees.

7.5.6.4 Control of root diseases

Control measures, which can be applied are similar for all root diseases. In small cacao farms it is advisable to clear-fell before planting and remove all stumps and bigger roots from the soil and establish temporary shade in the early years of the planting. When it is time to remove shade trees, tree killers should be used, alone or in combination with ring barking. This kills shade trees in a short time through poisoning and renders the area a less suitable medium for future fungus growth and development.

7.5.7 Twig and Leaf Diseases

7.5.7.1 Thread blight disease

This is caused by *Marasmius scandens* and other fungi. This disease can be recognized by the presence of dead leaves hanging down from a branch or twig of the tree. Closer observation reveals many thick white mycelial strands (strings) growing over the branches and the leaves thus killing them. Although the petioles of the leaves break off after having been killed, the leaves do not fall to the ground as they remain pendant on the mycelial strand.

This disease develops gradually but a large part of a tree canopy can be killed in time if the disease is not controlled. An efficient control method is to prune all infected parts and burn them on the spot. As the ends of the mycelial strands cannot be seen with the naked eye, pruning should be done at least 15 cm beyond the point where the mycelium ends.

In some cases the mycelial strands are not thick and white, but thin and black. The disease is then called horse hair blight, but control measures remain the same. It is caused by *Marasmius equicrinus*.

7.5.7.2 Pink disease

Pink disease is caused by *Corticium salmonicolor*. This fungus disease has the same growth habits as thread blight especially as it grows over twigs or branches

Initially it forms a continuous whitish film which later on turns pink. The twig (or branch) is killed after some time.

Spraying with Bordeaux mixture (1 per cent copper sulphate) or other copper fungicides helps to check the disease. It is also good to prune infected parts and burn them on the spot.

7.5.7.3 Calonectria die-back

This is a disease caused by a fungus *Calonectria rigidiuscula*. This fungus attacks only weakened trees such as those with wounds. It is always associated with damage done by mirids. The mirids weaken the trees by making punctures through which the fungus enters. The fungus then infects the surrounding tissue thus ringing the infected branch or twig and death of the abaxial part is then inevitable. Soon afterwards many whitish-pink cushions appear on the dead wood. These contain numerous spores of the fungus.

As vigorously growing trees are usually not infected, control of this disease can be achieved by controlling mirids. In addition it is advisable to prune spore-bearing parts and burn them on the spot.

7.5.8 Cherelle Wilt

This is one of the major diseases of cacao. The disease affects very young developing cocoa pods of about 10 cm in length or less. These suddenly wilt and die. Dead cherelles can be seen hanging on trees for long periods. The incidence of this disease often causes pod loss of about 40 to 50 per cent of the total produced by individual trees. This means that only about half of the developing pods on a tree actually reach maturity as a result of death of the rest caused by cherelle wilt. Unfortunately, the cause of this disease is not known at present, but it is believed to be the result of certain physiological changes taking place in the tree. Sometimes, it is suggested that the phenomenon is a method by which the tree regulates the number of pods it is able to carry and nurture to maturity, i.e. a physiological cherelle wilt. Other schools of thought have accused a number of fungi of causing cherelle wilt, and even insect pests. The different types of cherelle wilt in the field are still to be studied in detail.

Whatever the cause of cherelle wilt, it causes serious losses of cacao pods.

7.5.9 Cushion Gall

This is another disease of cacao of which the cause is not well known. One opinion holds that cushion gall is caused by *Calonectria rigidiuscula*; another that cushion gall is physiological disease which has a high degree of heritability. In cushion gall, the cushion which is the point of production of the cauliflorous flowers of cacao, is affected, the cushion produces a gall of flowers, the flowers drop and no pods are produced. Trees with cushion gall disease are easily identified on the field by massive clusters of living and dead flowers. Cushion gall is controlled by removal of the affected trees.

7.6 CACAO PESTS AND THEIR CONTROL

7.6.1 Some Major Pests

7.6.1.1 Cacao mirids

Mirids are the most serious pests of cacao in West Africa. They can kill young plants and badly damage mature trees thereby decreasing the yield of pods. It is important to recognize the mirids and their damage, and to know the best methods of spraying for lasting control.

Cacao mirids are also called capsids and 'jori-jori'. There species are found in West Africa.

1. The Brown Mirid (*Sahlbergella singularis*) is the common mirid, and attacks cacao in all parts of West Africa. It feeds on pods, chupons, and branches including the tips. The winged adults are also able to feed on the hard wood of branches several years old.
2. The Black Mirid (*Distantiella theobroma*) is found mainly in the Ife-Ondo-Akure area of Nigeria and only occasionally elsewhere. It feeds on pods, chupons and fresh green shoots, and especially on young plants.
3. The Cacao Mosquito (*Helopeltis bergrothi*) is rare. It feeds mainly on pods.

The mirids usually feed at night and rest during the day in the axils of the leaves, between touching branches and behind pods.

The Brown Mirid sometimes feeds on kola trees, and the Black Mirid is occasionally found on grapefruit, orange and silk cotton trees.

7.6.1.1a Life history

The Brown Mirid completes its life cycle in 40 to 52 days. The female lays eggs in the stems or pods of the cacao tree. The young insect (nymph) emerges after 13 to 18 days, and is only about 3 mm long. After feeding for about four days it is fully grown, its skin splits and the next stage nymph emerges. It grows in this way another four times, and the last time it develops into the winged adult. The adult females begin to lay eggs 4 to 11 days after mating. They can lay a total of about 200 eggs over a period of 30 days, and may live a few weeks afterwards. The adult males live for two or three weeks.

7.6.1.1b Mirid damage

Mirids have tubular mouthparts with which they pierce cacao and suck up the juices causing oval circular black patches of dying tissue, called lesions. After a few weeks the lesions on young green shoots become black sunken grooves, and lesions on hardened wood may become shallow oval pits. If young green shoots or small seedlings are ringed by lesions they may die, and repeated damage may kill young trees.

Often fungi (or mould) enter the cacao plant through the lesions and grow into the tissue immediately around the lesions, producing a patch of roughened bark (called a callus). The fungi can remain dormant for a long time, and then grow quickly when the tree is weakened, either by mirid attack, by swollen shoot virus, or during a dry period, sometimes several years after the fungi first entered the tree. When the tree is weakened these fungi may kill many of the small branches in the canopy and parts of the main branches, giving the tree a bad shape and reducing the yield for several seasons. These fungi probably do more damage than the mirids to mature cacao trees, but by killing the mirids the fungi cannot enter the trees easily.

The commonest type of damage is called 'die-back'. The branches or chupons die and the dead brown leaves remain attached to the tree. Die-back can also be caused by boring insects or thread blight. Therefore, a check should be made for the presence of holes in the bark or tunnels inside. If the die-back is caused by mirids there should be oval black scars or roughened bark on the dead branches. If there are pods on the tree, these should be examined for circular black spots, especially around the pod stalk, which will indicate that mirids have been on the trees recently.

When mirid die-back or mirid pod damage is found it is important to look for the mirids before spraying. Mirids usually inhabit the angels where branches meet, in leaf axils or behind pods, especially pods with black spots.

The nymphs are roughly oval in shape. The young nymphs are about 3 mm long and have an orange-red colour; the older nymphs are dark brown and up to 4 mm long. The adults have wings, and often fly to another cacao tree when disturbed; they are about 4 mm long, and either brown (the Brown Mirid) or black (the Black Mirid).

The nymphs of the Cacao Mosquito are slender and green, and have a spine on the back with a knob at the tip.

Heavy 'blast' damage is caused by the fungus *Calonectria rigidiuscula* associated with mirids. It starts to kill the branches when the cacao trees are weakened, either by mirid attack, by swollen shoot virus, or during a dry season.

If the insecticide spraying is done well, the mirid population will be reduced and cacao farms should give an increased yield, including those infected with virus in the area of mass infection (AMI). The effects of the dry season can be reduced if the farm is clean-weeded in October and January, so that more soil water is left for the cacao trees.

Mirid populations are generally lower in farms with proper shade. Young cacao trees should have thin overhead shade, which can be provided by forest trees or by planting tree cassava or plantains. The branches of cacao trees more than two or three years old should meet to form a continuous cover about 2 to 2.5 m above the ground.

7.6.1.1c Control

Gammalin 20 is the only insecticide recommended for use against cacao mirids. This is because it kills the mirids if used properly, and does not cause other cacao pests to increase, and the cacao beans are not tainted.

Gammalin 20 (100 ml) is diluted in 9 litres of filtered water, while stirring. Four litres of the mixture are sufficient for spraying 20 to 25 mature cacao trees.

It is recommended to use a sprayer with a 2 m long lance for large trees. The sprayer is pumped to 6 kg/cm^2 and, holding the nozzle about 1 m away from the tree, the spray is applied fairly quickly up the trunk and along one branch to the tip and then across the canopy to the other side; the tree is to be sprayed again from two more directions until every part of the tree is covered with the mist.

Mirid eggs are laid in the cacao tissue and are, therefore, not killed by the insecticide, so after one spraying it is necessary to spray again three to four weeks later when all the eggs have hatched.

Mirid damage is not very noticeable in the early rainy season when the trees are growing strongly but it is necessary to spray against mirids in August to prevent heavy damage later in the year. It is safest to spray three times starting in August, with intervals of three to four weeks.

After the third spray in October the farm should be inspected monthly in case adult mirids have flown in. If this occurs any trees found with mirids should again be sprayed.

Mirids tend to congregate in 'pockets', that is they occur on several trees close together. If mirids are found on one tree neighbouring trees should be inspected with extra care. Inspection for fresh damage and mirids should be a continuous process on the cacao farm. It is a waste of insecticide and money to spray in February or March even if the damage is severe because the dry weather will kill most of the mirids that are still alive by that time of the year.

Control measures are most efficient when large areas are sprayed at the same time. Where several farms of less than 2 ha are close together, control will be most efficient if all the farmers cooperate in spraying all the farms on the same day or, failing that, within one week.

Some insecticide including 'Gammalin 20' are now known to have become ineffective against the mirids in Ghana and in some parts of Nigeria, the mirids having developed resistance to such insecticides. This resistance may develop in other parts of West Africa and if a farmer reports that the mirids on his farm are not being killed by the Gammalin spraying, a check should be made to see if he is using the correct method. If Gammalin 20 has been sprayed thoroughly, then resistance is suspected and other recommended insecticides such as 'Elocron' could be used.

7.6.1.2 Earias

Earias biplaga belongs to the ENDOPTERYGOTA group of insects and it has four stages in its life cycle.

The eggs are laid singly usually just below a leaf bud. After five days, the first instar caterpillar emerges and burrows into the bud. After destroying one bud, it moves to other buds and destroys them, or eats some of the leaves. When there are several caterpillars on a young cacao plant all the buds may be repeatedly destroyed and the plant may die.

The fifth instar caterpillar spins a fibrous boat-shaped cocoon in which it turns into a pupa. The larval period takes about 14 days and the pupa takes 9 days before the small green adult moth emerges. The female moth lays eggs after mating, so that the length of the life cycle, that it is the period from one egg to another in the next generation, is four weeks. The eggs and pupae are not killed by the insecticide and the caterpillars are boring in the buds for at least half of their lifetime. Their most vulnerable period is when the young caterpillar hatches. Therefore, a double spraying five days apart is recommended; the first spray kills any exposed caterpillars while the deposit will kill any caterpillars moving from one bud to another during the next few days and also young caterpillars emerging from the eggs. The second spray is necessary to protect buds which have grown since the first spray. 'Rogor 40' is a very effective insecticide. Spraying may be necessary when the seedlings are in the nursery, and at the beginning of the rains, when the shade is first thinned to harden off the seedlings. When seedlings are transplanted to the field, it is necessary to inspect the buds of a few plants and give a double spray if the buds turn brown and die.

Earias is scarce in the dry season but it is a commom pest during the rainy season. If nurse shade is well maintained *Earias* attack will be reduced. *Earias* becomes a minor pest after the cacao canopy has been closed.

7.6.1.3 Mealybugs

Swollen shoot disease of cacao is caused by a virus, which has to pass through living mealybugs to get from one cacao tree to another. The mealybug is therefore known as the **vector** of the virus.

7.6.1.3a Life history

Mealybugs are the only insects so far known to be able to transmit the cacao swollen shoot virus. They are plant-feeding bugs (**Hemiptera**) with long and flexible mouthparts which can penetrate deeply into the plant's tissue. When not in use these mouthparts (stylets) are coiled up in a sac within the insect's body.

The life history of mealybugs is unusual. The females develop from eggs through three nymphal stages, the last leading directly to the adult female which is wingless and looks much like a nymph except that it is bigger. The adult males, on the other hand, develop through what is called a prepupa and are winged, lack mouthparts and do not feed. They are short lived and rare so that in many species, females reproduce without males. A female may lay from 30 to 500 eggs according to the species involved.

Up to 90 per cent of the movement in mealybugs occurs in the first nymphal stage which is, therefore, the most important stage in the spread of virus from tree to tree. This spread is achieved mainly in two ways – by nymphs moving from one infected tree on to a healthy tree and feeding on it when the branches touch, and by the small, light nymphs being carried by the wind from infected to healthy tree. It is unlikely that nymphs move across the ground from one tree to another in any numbers, and transport by ants has been demonstrated but probably does not account for much of the virus spread.

At least a dozen different species of mealybug infest cacao in West Africa. Most are uncommon and the bulk of the population is made up of three species:

Planococcoides (= *Pseudococcus*) *njalensis*
Planococcus (= *Pseudococcus*) *citri*
Ferrisiana virgata

P. njalensis occurs only in West Africa but *P. citri* is found in many parts of the world. The two species are thought to be equally important in West Africa. The third species *Ferrisiana virgata* is also found often on seedlings, flowers and flower cushions where it probably causes some flower fall.

7.6.1.3b Mealybugs and the virus

The virus found in cacao is a mixture of strains or **isolates**. Each isolate is recognized by its effect on the plant. Special microscopic techniques have shown that isolates may also differ in shape. Not all mealybugs can transmit all isolates equally or at all. A mealybug may acquire enough virus for transmission after only 1 hour's feeding on an infected plant or after up to 10 hours, depending on the virus strain and the mealybug species involved. The transmission of virus by a mealybug may be as quick as 15 minutes feeding time on a healthy plant or as long as four hours. The persistence of virus in a mealybug is greater when the insect is starved. Occasionally a mealybug may transmit virus after 36 hours' starvation but in field conditions the insect would normally not starve for that long and it appears that the virus does not then persist for more than 2 hour.

7.6.1.3c Mealybug control

Attemps made to control mealybugs can be divided into two categories: **biological** and **chemical**.

Biological control has been attempted by releasing mealybug parasites into the field in large numbers but the results were not promising.

Chemical control by spraying mealybugs has not succeeded because the waxy coating of the insects prevents contact insecticides from reaching their bodies. Systemic insecticides which pass through the plant and could be sucked by the mealybugs often have side effects like toxicity to the cacao plant and the presence of unpleasant tastes in the chocolate manufactured from beans harvested from treated trees.

One method of chemical control has been directed at the ants which often protect mealybug colonies with debris in return for the honeydew of the mealybugs. When a persistent insecticide like 'Aldrin' is used against ants their numbers fall rapidly and soon after, the mealybugs, being exposed, also fall sharply in numbers. But soon there is trouble with minor pests which had hitherto been controlled by various parasites and predators, including the ants.

Currently in Ghana a method of controlling mealybugs with a fungus is being investigated. A fungus (*Cephalosporium* sp.) is cultured in rice crushed in water and then sprayed on mealybugs which are killed within a few days. This method is yet too expensive for use on cacao plantations, but may be developed into a practical control measure in the future.

7.6.2 Insect Pests of Young Cacao in the Nursery

To obtain strong healthy seedlings, cacao beans should be sown in a good forest soil in black polythene pots in a shaded nursery and watered every other day. Damage by insects can be easily detected and controlled as follows:

1. If many holes are found in the leaves, the underside of all the leaves on the damaged plants are checked; any caterpillars and beetles should be killed. If there are many noxious insects, it may be easier to spray the plants with DDT. If no insects can be found, damage may be caused by beetles or grasshoppers that feed at night and hide during the day. This sort of pest can also be controlled with several insecticides.
2. If the buds turn brown and die and contain small, brown, spiny caterpillars (the spiny bud worms) all the seedlings should be sprayed with Rogor. This insecticide will also kill three other pests of the buds: small white mealybugs, colonies of small black aphids and small psyllids. The psyllids have a cover of white threads like cotton-wool.
3. If the leaves become pale green in patches and have creamy white particles like fine sawdust on the under surface, or if the leaves have many brown spots and fall, very small black or yellow and red insects on the underside of the leaves may be found. These are spider mites or thrips. they can be killed with Rogor.
4. When the leaves wilt, turn brown and stay on the seedlings, the roots should be inspected. If the main root has been eaten away and there are many small, white insects present, these are termites and need to be sprayed with Aldrin or Dieldrin.

7.6.3 Insect Pests of Young Cacao in a Farm (see Figure 24)

When cacao leaves wilt and die, especially in the dry season, examine the roots. If the main root is eaten away and several small, white insects are found in the holes these are *termites*. These can be controlled with Aldrin or Dieldrin. Wilting can also be caused by root fungi, or drought. When leaves wilt and there is a tunnel in the centre of the trunk or branch a *borer* is responsible.

1. If there is a single large hole in the bark of the trunk or branch exuding watery sap or gum, and the tunnel inside is long, the damage is done by the trunk borer (*Eulophonotus* sp.).
2. If in the dry season the hole is narrow and the tunnel is only about 10 cm long the damage may be made by a beetle, (*Apate monachus*).
3. If there is a row of small holes along the dying branch with sap exuding or sawdust protruding from the holes, the damage is made by a stem borer (*Tragocephala* sp.).
4. If the hole is as small as a pin head and sap is exuding, or there is a white

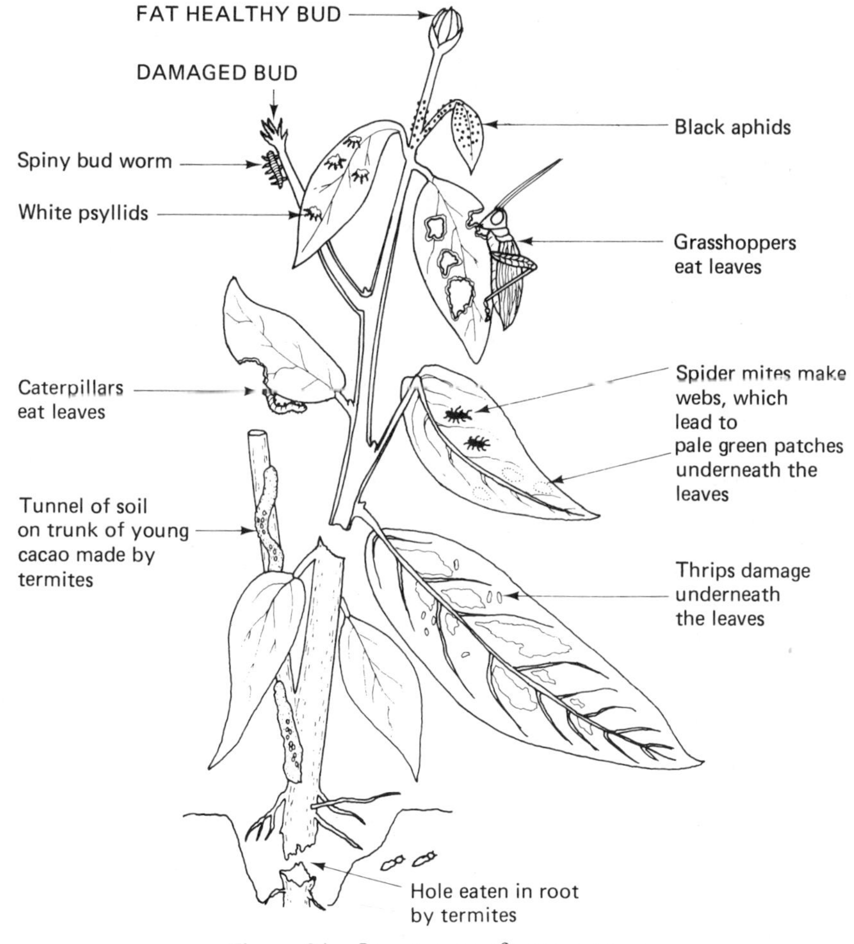

Figure 24. Insect pests of young cacao

streak of dried sap below the hole, the damage is made by the shot-hole borer (*Xyleborus* sp.).

Die-back damage without holes and tunnels in the branches, but with oval black scars on the surface is typical of recent attacks by mirids. When there are oval pits and roughened areas of bark only look for the mirid nymphs in the angles where branches meet or where leaves join the branches.

If cacao leaves are being eaten up, it suggests caterpillar damage. Noctuids are the most common leaf-eaters, but arctiids, lymantriids and sphingids are also common.

If the buds are turning brown and dying the spiny bud worm, *Earias*, is responsible. If buds are damaged repeatedly the young cacao may be killed so it is important to be able to recognize *Earias* attack.

7.6.4 Insect Pests of Mature Cacao Trees

Borers can be found in the trunks (trunk borer, *Eulophonotus* sp.) and in the branches (stem borer, *Tragocephala* sp., and shot-hole borer, *Xyleborus* sp.). The die-back damage is the same as described for young cacao (see 7.6.3).

If trees have lost leaves from several of the branches, and small brown or black spots occur on leaves remaining on the affected branches particularly on the underside where the veins join the midrib, thrips damage is likely. If the pods also have a bronze discoloration either in the furrows, or over the whole surface, a typical symptom of *thrips* damage (*Selenothrips*) prevails.

Die-back of branches or chupons without holes in the bark or tunnels inside, suggests mirid damage, characterized by oval black scars or roughened bark on the affected branches, and circular black spots on pods. This can be confirmed by finding the mirid nymphs in the angles where branches meet, in leaf axils or under pods.

Pod damage occurs in the following forms:

1. Sometimes a black mass of frass, or faeces, held together by silk threads is found on a pod where it touches the tree trunk, usually near the pod stalk. When the frass is removed a pale pink, or pale brown caterpillar of *Characoma* (Noctuidae) is seen eating tunnels in the pod husk. The caterpillar usually does not penetrate the husk to destroy the beans so it is not an important pest.
2. The pod-husk miner is a small caterpillar that eats just under the surface of the pod leaving a roughly circular blotch of black skin about the size of a penny. When the blotch is brown the caterpillar is still feeding inside, but when the blotch becomes black the pod is usually empty and the skin can be rubbed off easily. The miner is not a serious pest.
3. If the furrows or the whole surface of the pod have a bronzed colour, and the green or yellow colour of the pod is easily seen when the surface is scratched, the brown deposit has been made by thrips (*Selenothrips*).

4. The common cacao mirids (*Sahlbergella singularis* and *Distantiella theobroma*) make circular black feeding marks on pods, especially around the pod stalk. When several nymphs have been feeding most of the pod at the stalk end becomes blackened, particularly on the side nearest the trunk.
5. The rare cacao mirid *Helopeltis bergrothi*, sometimes called the Cocoa Mosquito, makes smaller circular black feeding marks on pods, and these are found on all parts of the pod.
6. In 1964 a coreid pest, *Pseudotheraptus*, was discovered on cacao pods in Ghana. It feeds on guava and avocado pear and makes black spots on these fruits in Ghana and in Nigeria. It is feared that this insect may start feeding on cacao pods all over West Africa. The damage to the pods is severe; many young pods are killed and mature damaged pods are often less than half the normal size, with deep depressions all over the pod surface containing black circular feeding marks often with a white streak of dried sap on and below the black mark. The adult is a pale brown insect, longer and thinner than a mirid, and with long antennae. The old nymphs are brown and the young numphs are green and brown, and their antennae are very long with a black knob near the tip.
7. *Earias* sometimes eats into pods, which usually become distorted with a depression on the side of the pod where the whole is formed.
8. When a pod is thinner than normal and is distorted, especially if the distortion is like a waist around the middle, and if there are no holes in the pod, the damage may be that of *Bathycoelia*. *Bathycoelia* nymphs are green with small black spots, and red legs and antennae, and are usually found on the pods. The adults are dark green with yellow spots down the sides, and two yellow spots in the corners of the triangular shield on their backs.

7.6.5 Control Measures

Three main types of control measures are possible against cacao pests: cultural, natural and chemical control.

7.6.5.1 Cultural control

This consists of appropriate farm management practices and adequate farm hygiene.

Shade is an important cultural method. Young cacao in a nursery should be shaded. When establishing a farm on land without natural shade, young cacao seedlings should be shaded with plantain. If a new cacao farm is being established some of the smaller trees should be left for shade. When a farm is well maintained the canopies of the cacao trees should close after two or three years.

One important reason for providing shade is that attack by many insect species is reduced by adequate shading. These species include mirids, *Earias*, *Anomis*, thrips, and some of the borers. However, if the shade is too heavy growth of the cacao trees will be adversely affected. A good farmer will examine his cacao trees every month and remove dead branches. If holes made by trunk borers are spotted one of the best methods of control is to poke a thin wire into the tunnel and thus kill the insect.

Weeding is important because weeds compete with cacao trees for nutrients and water in the soil, and many weeds are hosts for cacao pests, for example *Earias* feeds on *Sida stipulata* which is a common weed.

7.6.5.2 Natural (biological) control

Many insects are controlled to a certain extent by predators, parasites or disease organisms. The main insect predators known in cacao are mantids, reduviids and ants; the main parasites are wasps. Sometimes the natural controlling agents are not efficient and insects such as mirids and *Earias* reach pest status on cacao. It then becomes necessary to control such pests by other methods. Cultural control methods have the advantage of cheapness, while natural control is usually free.

7.6.5.3 Chemical (insecticide) control

Chemical methods of control are expensive, but they give a quick control when used properly. It is most important to follow the recommendations given by experts because those are designed to give the best control as cheaply as possible. Insecticides for use on cacao must be tested to ensure the following conditions:

1. The insecticides do not damage the cacao trees, particularly the young leaves.
2. When used as recommended the insecticides do not harm or poison the farmers.
3. The chocolate made from cacao beans harvested on sprayed farms does not have an unpleasant taste and is not poisonous.
4. The insecticides do not cause minor pests to increase.
5. The insecticides do not prevent pollination of cacao flowers.
6. The insecticides are the cheapest and most efficient when used at the appropriate concentration, methods, and time of year for application.

For mirid control 'Gammalin 20' (or lindane) is recommended. 'Aldrex 40' is not recommended because it is known to cause an increase in other pests. This is probably because it is persistent, and remains on the trees for several weeks killing predators and parasites of insects which are inside the cacao trees and are not exposed to the insecticide.

Therefore, branch borers, pod-husk miners and *Characoma* survive the spraying while their natural controlling agents are killed with the results that these minor pests are allowed to increase without restraint. If a farm has a lot of borer damage and about half the pods have *Characoma* caterpillars and several

Table 9 Summary of insect pests of cacao

Common name	Scientific name	Family	Pest status	Damage done	Control measures
Mirid/capsid	*Sahlbergella singularis* *Distantiella theobroma*	Miridae	Major	Die-back on shoots. Stag headed cocoa, lesions on pods	Spray with Gammalin 20 where there is no resistance to lindane. Spray with Elocron or Unden where insects show resistance to lindane
Cacao mosquito	*Helopeltis bergrothi*		Minor	Black spots on surface of pods	
Cacao shield bug	*Bathycoelia thalassoma*	Pentatomidae	major/minor	Suck terminal or distal end of cherelles. Cause yellowing and wilting of cherelles	The use of Elocron is proposed
Leaf caterpillars	Anomis ieona	Noctuidae	Minor/major	Defoliate leaves. May eat into cherelles	Spray with Gammalin 20, Elocron or Rogor
Defoliators	*Earias biplaga* *Eariasinsulana*	Noctuidae	Minor/major	Eat terminal buds and young leaves	
	Laphocrama sp.	Noctuidae	Minor	Eat leaves and epidermis of pods	
Leaf worms	*Spodoptera littoralis*	Noctuidae	Minor	Feed on seedling and leaf flush	Spray with Gammalin 20 or Elocron or Rogor
Leaf defoliator	*Anaphe venata*	Notodontidae	Minor	Larvae are gregarious and eat leaves	Spray with Gammalin 20, Elocron or Rogor
Pod-husk miner	*Marmara* sp.	Lethecollectidae	Minor	Mine into the pod's epidermis. Discolour pods	

(*Continued*)

Table 9 (*Contd.*)

Common name	Scientific name	Family	Pest status	Damage done	Control measures
Pod-husk borer	*Characoma stictigrapta*	Noctuidae	Minor/major	Eat pods and sometimes beans; may eat leaves	Spray with Gammalin 20
Bag worm	*Kotochalia* sp.	Psychidae	Minor	Eat leaves and epidermis of pods; use branches as tents	
Variegated locust	*Zonocerus variegatus*	Acrididae	Minor	Eat leaves	Gammalin 20, Unden 20
Termites	*Nasiutitermis* sp.	Formicidae	Minor/major	Eat branches or roots	Spray with Aldrex 40 or Agrothion
Tailor ants	*Oecophylla longinoda*	Formicidae	Minor	May tie leaves of plant to form nests	Spray with Rogor or Sevin
Longhorn beetle	*Tragocephala castnia*	Cerambycidae	Minor	Bore into branches and stems	Poke larvae in tunnel, squash adult. Spray plant with Sevin
Shot hole borer	*Xyleborus compactus*	Scolytidae	Minor	Come into twigs and seedlings	Spray plant with Sevin
Red banded thrips	*Selenothrips rubrocinctus*	Thripidae	Minor	Scrape epidermis of leaves. Cause wilting. May aggravate mirid damage	Spray with Rogor 40
Mealybugs	*Planococcoides njalensis* (Lanj) *Planococcus citri* *Ferrisiana virgata*	Pseudococcidae	Minor	Piercing and sucking injury reduce plant growth and crop yield	
Scale insects	*Stictococcus sjostedti* (Ck11.)	Coccidae	Minor	Feeding causes stunted growth	

Black citrus aphids	*Toxoptera aurantii*	Aphidae	Minor	Reduce plant vigour	Spray with Rogor
Leaf beetle	*Ootheca mutabilis*	Chrysomelidae	Minor	Eat leaves. Punch holes in leaves	
Spider mites	*Tetranychus* sp.	Arachnidae	Minor	Scrape underside of leaves in nursery	Spray with Rogor
Ants	*Oecophylla longinoda*	Formicidae	Minor		
Ants	*Pheidole megacephala*	Formicidae	Minor		
Ants	*Campanotus* sp.	Formicidae	Minor	They bite and sting farmers	Use of chemicals to block them
Ants	*Crematogaster gabonensis*	Formicidae	Minor	They tend scale insects or aphids. Knit some leaves together in making tents	
Ants	*Crematogaster* sp.	Formicidae	Minor	Some use soil or other plant debris to make tent on cacao; they have been associated with black pod disease	
Ants	*Macromiscoides aculealus*	Formicidae	Minor		
Ants	*Acantholepsis*	Formicidae	Minor		

pod-husk miners it is fairly certain that the farmers has been using the wrong insecticide, or possibly too much 'Gammalin 20' and too often.

7.7 A SUMMARY OF INSECT PESTS ON CACAO

Various insect pests of cacao are listed in Table 9.

7.8 CACAO HARVESTING AND PROCESSING

7.8.1 Harvesting Procedures

It takes cacao pods from 110 to 130 days, depending on variety, from pollination to pod ripening. To ensure the production of quality beans, it is essential that only mature and ripe pods are harvested and processed promptly. Diseased or damaged pods should never be processed for the market. In West Africa, there are two pod production seasons; the main season – July to December and the light crop season – January to April. Harvesting should be regular and frequent. This prevents not only overripening but also minimizes losses due to diseases and animals. For this purpose, harvesting should be done twice a month for the main crop season and at least once a month for the light crop season. Tools required for pod harvesting include the following:

1. Sharp cutlass for plucking those pods which are within easy reach on the cacao trees.
2. Harvesting knife with short handle for harvesting ripe pods well above ground level.
3. Harvesting knife attached to long poles for plucking pods from the topmost part of cacao trees.
4. Basket, or any convenient container for packing the harvested cocoa pods from the plot to the place where the beans are to be separated.

It is important that harvesting knives are very sharp so as to avoid damage to the bark of the trees. Bad harvesting techniques reduce yield of cocoa in subsequent years. Therefore, while harvesting, the following points should be noted:

1. Cacao trees should not be climbed to avoid rubbing off those flower cushions which are the source of future pods.
2. However low a pod may be on the tree, it must not be pulled off by hand because this may result in damage to the cushions providing a hiding place for insect pests and pathogens.
3. Flowers borne on top branches usually fall off more easily. To prevent loss of such flowers, the branches as much as possible should not be shaken vigorously during harvesting.
4. Generally, any wound to the bark of the cacao may serve as an entrance

for fungal infection. Therefore, great care must be taken to avoid wounding when harvesting.

5. Diseased pods or those damaged by rodents should be removed and buried or placed far away from the cacao plot.
6. After harvesting, all good fermentable pods should be taken to the processing area without delay.
7. Pods with initial stage of black pod infection are generally found to be good and their beans fermentable.
8. It is absolutely necessary to break the pods for fermentation soon after harvesting in order to prevent the onset of beans germinating within pods or fermentation which will be incomplete.

7.8.2 Breaking the Pods

The pods are broken by knocking them against blunt objects, e.g. a thick piece of wood or stone. The beans and the pulp are removed from the pods. This method has been found to be quick and efficient. A man may break over 3000 pods a day. Pods should not be broken with a cutlass as this often results in damage to the beans which are then no longer suitable for fermentation.

The extracted beans are collected in a clean container, usually a basket, care being taken not to include beans of the following categories:

1. Those which have started to germinate.
2. Beans which are too small because they are immature.
3. Beans looking dry or diseased in any way.
4. Beans from damaged pods.
5. Beans from black or diseased pods.

All the undesirable forcign elements including pulp and pieces of broken husk should be removed from the freshly extracted beans before fermentation.

7.8.3 Fermentation

Fermentation is one of the most important operations in the preparation of cocoa beans for the market. As a rule, maximum return is obtained only from properly fermented and clean cocoa. It is necessary to ferment cocoa beans in order to fulfil the following conditions:

1. Get the proper taste, colour and flavour associated with cocoa products.
2. Kill the embryo and stop germination.
3. Remove pulp or mucilage or sweatens so that the beans may dry properly.
4. Loosen the skin from the cotyledon thereby allowing easy and proper de-shelling during processing.

7.8.3.1 Heap fermentation

Straight pieces of wood about 2 to 3 m long are arranged about 15 cm apart on clean ground to form a base. Banana or plantain leaves are spread over the prepared platform. These leaves are perforated in a number of places to allow for the drainage of sweatings. Wet beans are then piled on this base until a heap 60 to 90 cm high is formed. The heap is then covered with banana leaves held down with logs of wood or pieces of banana stem. The proper covering of the heap helps to retain its heat.

On the third and fifth days the heap is thoroughly turned and evenly mixed with a wooden spade or by hand. This should be done as quickly as possible to avoid heat losses before the heap is covered up again. If properly done, the cocoa beans should be fully fermented and ready for drying on the seventh day. Although this method is common and has replaced pit fermentation, its disadvantage lies in the problem of turning, the long period of fermentation and its unsuitability for small quantities of cocoa.

7.8.3.2 Basket fermentation

Woven baskets, preferably of medium size, which can be easily lifted, are used. The baskets are lined with perforated banana or plantain leaves to facilitate drainage of sweatings. The baskets are also placed on a wooden raised platform to aid further drainage. After being filled, the baskets are tightly covered with leaves held down with logs or stones in order to minimize heat losses and encourage the flow of effluent.

On the third day, the baskets are emptied into newly prepared ones and covered as before. This turning of the beans from one basket to the other should also be done as quickly as possible to avoid the loss of heat. On the fifth day this process of transfer should be repeated as before.

On the seventh day, the cocoa beans should be fully fermented and ready for drying. There is no reason why this method of fermenting cocoa should fail to give good results if done properly. Its disadvantages are similar to those discussed under heap fermentation.

7.8.3.3 Sweat box fermentation

This is an improved method of fermentation employed mostly in large commercial plantations. The method is also practicable for large-scale farmers who can make some capital outlay available. It has been widely used for many years and found satisfactory by large cacao estates in West Africa.

Three wooden boxes made of hard wood are used. The bottoms of the boxes are perforated to facilitate drainage. Each box is $90 \times 90 \times 90$cm, but $120 \times 90 \times 90$ can be used. No lids are necessary, one side of each box should be removable. Nails, if used to build the boxes, should not be allowed to make contact with the beans as this can cause tainting. The boxes are put on a raised platform for aeration and free drainage. This method requires a ventilated and roofed building to protect the boxes from rain. Such building should also be provided with an adequate drainage system for the effluent. After the boxes have been arranged in steps one after the other the wet cocoa beans are put in the topmost box and covered with sufficient banana or plantain leaves and sacks. The covering needs to be weighed down with logs of wood or stone to facilitate fermentation. Beans from the topmost box are turned into the middle box on the third day and covered as before. On the fifth day, beans are again turned from the middle box into the bottom box (i.e. the last box) and covered as usual. During each turning from one box into the other the beans should be quickly mixed (Figure 25). On the seventh day, the beans are fully fermented and ready for drying. Fermentation may be delayed if beans are not properly turned and handled when emptying them from one box into the other. Its disadvantages are similar to those discussed under the heap fermentation.

7.8.3.4 Tray fermentation

This method is used in most cocoa producing countries. It is a fast and an efficient method of fermenting cocoa which is now the standard practice at a number of cacao estates in Nigeria.

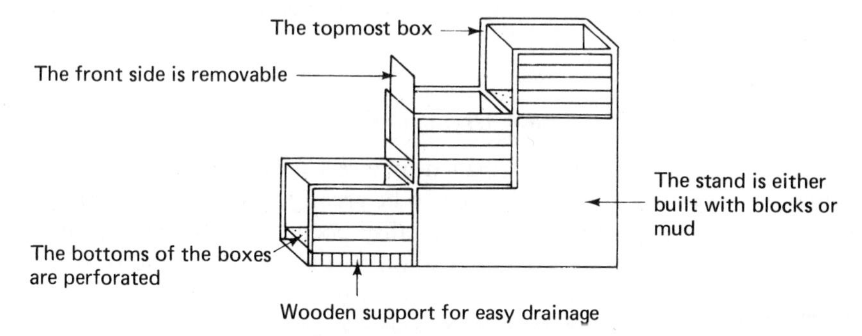

Sweat box fermentation method

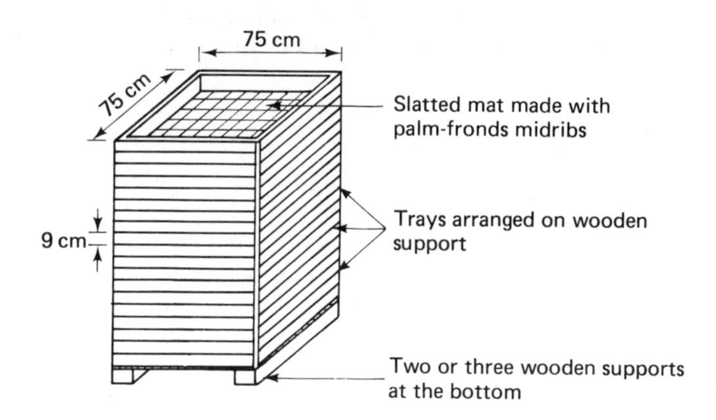

Tray fermentation method

Figure 25. Sweat box/tray fermentation methods

Strong woods are used to construct the four sides of each tray but the bottom is covered with slatted mat made from palm frond midribs. A small tray measures 75 × 75 × 10 cm.

After the extraction of the beans from the pods the trays are filled to the required level with wet beans. The first tray is put on a raised wooden platform to facilitate drainage and aeration. The volume of the harvest will dictate the number of trays to be used. But 10 or more trays could be stacked one upon the other to a convenient height for handling. It is very important to cover the top tray with banana or plantain leaves and sacks to retain the heat. Drying starts after the fourth day of commencing fermentation. The advantages of this method are as follows:

1. It requires far less labour to handle as compared with other methods. The average cost of the tray used is low and a tray could be used for more than two years if carefully handled.
2. It is fast, as only four days are required to complete fermentation.
3. No mixing of beans is required and well-fermented good quality beans are obtained.

7.8.4 Drying of Cacao

Stored cacao is often seriously damaged by insect pests and most especially by moulds. This type of damage is known to be greatly favoured by failure to dry cocoa properly. Production of good quality cocoa will, therefore, not only depend on proper fermentation, but also on correct drying methods. Moisture content should be 7 per cent, the range being 6 to 8 per cent.

7.8.4.1 Methods of drying cocoa

Various methods of drying the fermented cocoa beans are used. These are classified into sun drying an artificial drying.

7.8.4.1a Sun drying

The climatic conditions of a cocoa producing country will dictate the method of drying to be adopted. In West Africa, for example, cocoa beans are generally sun dried. This is possible because the bulk of the main crop is produced towards the beginning of the dry season which gives favourable conditions for drying. Sun drying is simple, cheap and very effective for most farmers who usually have small cocoa holdings. Drying is done on concrete slabs or raised platforms.

A reasonably open and flat site is chosen on which to construct a concrete slab. The size of the area will depend on the amount of produce to be handled and, of course, on the financial ability of the individual concerned. The area chosen should be well cemented with adequate drainage round it. For local farmers who keep livestock, the platform sould be properly fenced with fine wire netting to keep away all classes of livestock and prevent contamination of the cacao beans.

The slab should always be kept clean. After being fully fermented, the beans are spread as thinly as possible for sun drying. During the process of drying, all foreign matters and deformed beans must be picked. Occasional hand rubbing of the bean mass is essential for uniform and satisfactory drying. It is also very important to separate, by hand, all those beans sticking together. This will also promote even drying of the beans.

An alternative method is to construct a raised platform. Materials required for this can be easily obtained around the farm. Bamboo posts are used to erect the platform. The height could be 90 to 100 cm or as convenient. Slatted mats are then spread on the platform for drying cacao beans. Drying cacao by this method is equally good and effective if properly handled. In good sunny weather, it takes less than two weeks to achieve a satisfactorily low level of moisture in the beans.

7.8.4.1b Artificial drying

The relatively low rainfall during the cropping season in most cacao growing countries, allows for sun drying; however, in many other countries where the conditions do not favour sun drying, artificial methods must be employed. Artificial drying is employed in Cameroun, Costa Rica, Fernando Po, Surinam, Brazil, Panama and Malaya. These countries dry their cocoa beans with different types of artificial driers.

In West Africa many large plantations which have recently been planted with new varieties of cacao, are now producing large quantities of ripe pods throughout the year including the periods of very heavy rainfall. Very serious consideration will, therefore, have to be given to the large-scale use of artificial methods of drying.

During the heavy rains, other methods used on a small scale are the 'Evepy' plastic drying shed; trays on a pulley system; and the Cameroun drier – widely used and cheap.

7.8.4.1c Methods of artificial drying.

1. By spreading transparent polythene sheeting at some distance above thinly spread cacao beans rain can be prevented from reaching the drying beans, while part of the available radiation still serves to promote the drying process. It is essential that sufficient circulation of air is allowed to remove any moisture. It is usually necessary to dry finally the cacao beans by direct exposure to the sun in order to reach a satisfactory low level of moisture (8 to 10 per cent).
2. If sufficiently long periods of sunshine occur a temporary protection against any rain would serve to allow the drying process to proceed. Such

protection can be provided by hinging covers or roofs, which are raised in a vertical position during rain-free periods and lowered when rain threatens. Another system provides wheeled drying trays which are rolled under a fixed roof whenever rains are expected.

3. In the absence of long sunny periods artificial heating may have to be used. An example of this is the 'Cameroun drier'. This consists of a flue through which heat is passed to a chimney, by burning firewood. The flues pass through a well-ventilated construction, well protected against rain, and are covered with pebbles (3 to 4 cm) and fine sand (2 cm), on top of which mats are placed. Heat is thus dispersed and helps to dry speedily any cacao beans spread on the drying mats. Smoke from the heat source should not be allowed to contaminate the beans.

7.8.4.2 Test for dryness

Before storing cocoa, it is necessary to ensure that the beans are properly dried. Well-dried beans will crack when squeezed between the fingers. Another approach which is more accurate is to cut through the sampled beans with a knife and, if properly dry, the cotyledons will separate easily. This method is also used for testing other defects in cocoa.

7.8.5 Storage

When cocoa beans are properly dried and defective beans and other foreign matters picked out from the bulk, they should be put in clean baskets or new bags before being sold.

Cocoa is a delicate and sensitive crop which can readily absorb foreign flavours. Much good cocoa could thus be spoilt through careless handling. Damage can be avoided if the following rules are strictly observed:

1. Store the dry beans in clean baskets or new sacks and make sure that the baskets or the sacks are kept off the ground and walls, in weather-proof surroundings. Old sacks which have been used for storing maize or other foodstuffs should not be used as they are likely to harbour weevils and other insects which may attack cocoa.
2. Cocoa should not be stored near maize, tobacco, or other foodstuffs and should be kept away from smoke as this will produce smoky beans.
3. The produce should be sold to the buyers or export firms as soon as possible. Prices are fixed for the whole season and there is no advantage in keeping cocoa in the house where it is liable to deteriorate in quality.

7.8.6 Grading

In West Africa, there are two cropping seasons, namely the main crop which obtains between August and December and the light crop between January and April. To grade cocoa, representative samples are taken at random from the bulk. Several hundred such beans are taken but only 300 beans are finally selected and weighed. The 300 beans taken from the main crop should not weigh less than 300 g. The weight as stated may be affected by the variety of cocoa involved.

There are two grades of cocoa based on the following criteria:

Grade I cocoa: Less than 3 per cent slaty beans
Less than 3 per cent mouldy beans
Less than 3 per cent other defectives.

Grade II cocoa: Less than 5 per cent slaty beans
Less than 4 per cent mouldy beans
Less than 5 per cent other defectives.

Table 10 Types of unacceptable defective cacao beans

Defective beans	Cause	Remarks
Smoky beans	Storing beans in smoky area or room or near fire. Drying over open fire or dryer	Beans not acceptable
Velvety beans	Leaving ripe pods on the tree unharvested for several weeks, collected beans from pods damaged by rodents. Collecting beans from black pod diseased pods	Beans not acceptable
Black beans	Not fermenting in time. Drying on iron or metal sheets. Collecting beans from black pod disease	Beans not acceptable
Mouldy beans	Under-dried and bad storage. Storage in a damp place	Reduces the price and renders the cocoa unsaleable if found in large numbers
Weeviled beans	Collecting beans from pods damaged by rodents. Storing cocoa near tobacco, maize and other foodstuffs. Storing in a damp place. Leaving bags on the ground	Reduces the price and renders the cocoa unsaleable if found in large numbers
Flat beans Empty beans	Beans from immature pods in which the cotyledons are not well filled. They are not diseased	Lowers the price and makes the cocoa unsaleable if found in large numbers
Germinated beans	Leaving ripe pods on trees unharvested for several weeks. Fermenting in holes in the ground. Not turning beans during fermentation	Lowers quality and reduces price. Cocoa unsaleable if too much is found
Slaty beans	Result of non-fermentation of beans	Beans not acceptable
Purple beans	Result of under-fermentation	Lowers the quality and reduces the price. Unsaleable if too many

Certain kinds of cocoa beans are forbidden entirely and other kinds, if present in the cocoa, will either result in a lower price being paid for the cocoa or if there are too many of them, may render the cocoa unsaleable. The list of such defective beans with their causes is presented in Table 10.

7.9 USES OF CACAO

The Aztecs were the first people to use cocoa as food. They apparently made a fermented liquor from the beans and also a customary beverage as food. The beans, probably fermented in the pods and sun dried, were roasted in earthen pots and dehusked by hand. The nibs were ground on a slightly concave stone which at times was heated. After grinding, vanilla, spices, herbs of various kinds were added and ground in along with maize if a milder taste or lower cost was desired. The maize was preferably fermented before mixing with the otherwise finished chocolate beverage.

The use of cocoa as a drink or food of a very dietetic value by the Aztecs was not based on any scientific research. Things have changed. Today a great part of the world population knows about this 'food of the gods'. Science has added to our knowledge of cocoa and has thereby laid a foundation for its expanded usage.

The cocoa pod is composed of about 42 per cent beans, 2 per cent sweatens (mucilage) and 56 per cent husk. The bean itself is composed of the nibs (cotyledons and embryo) and the testa, the testa being about 1 per cent of the bean.

7.9.1 The Cocoa Bean and its Composition

The chemical composition of the most important part of the pod, the cocoa beans, is given in Table 11.

Cacao is free from any acutely toxic components. The nitrogen component contains two alkaloids – theobromine and caffeine. These two alkaloids are water soluble. Theobromine is present in fat-free dry cacao solids to about 3.5 per cent. It is a stimulant of muscular activity. Harmful symptoms such as excessive stimulation of kidneys, heart and smooth muscles has been associated with high consumption of theobromine. Caffeine increases sensory impressions. The small quantities consumed in chocolate have been proved harmless to man. The presence of these alkaloids in the cacao bean has made the feeding of unmarketable cacao beans to livestock on a large scale inadvisable. Attention must also be called to the high tannin and negligible acid contents of the cacao beans.

The cacao bean contains about 50 per cent fat. Since modern processing was embarked upon, cacao beans have served as a valuable source of vegetable fat – the cocoa butter. The residual cocoa powder is used in cakes, biscuits, Bournvita, Pronto, Coacfood, cocoa mixes, other confectioneries, ice cream, baking, dairy, soft drinks and in pharmaceutical and food preparations.

Table 11 Chemical composition of the cotyledons of fermented and unfermented cacao beans after drying

	Unfermented cotyledons		Fermented cotyledons	
	per cent in original	Per cent in fat-free material	Per cent in original	Per cent in fat-free material
Water	3.65	—	2.13	—
Fat	53.05	—	54.68	—
Ash:				
Total	2.63	6.07	2.74	6.34
Water-soluble	1.02	2.63	1.25	2.89
Water-insoluble	1.61	3.72	1.49	3.45
Sand, etc.	0.05	0.12	0.05	0.11
Alkalinity as K_2O	0.51	1.18	0.41	0.95
Chloride	0.012	0.028	0.014	0.032
Iron (Fe_2O_3)	0.004	0.009	0.007	0.016
Phosphoric Acid (P_2O_5)	0.96	2.22	0.60	1.39
Copper	0.0024	0.0055	0.0028	0.0065
Nitrogen:				
Total	2.28	5.27	2.16	5.00
Protein nitrogen	1.50	3.46	1.34	3.10
Ammonia nitrogen	0.028	0.065	0.042	0.097
Amide nitrogen	0.188	0.434	0.336	0.778
Theobromine	1.71	3.95	1.42	3.29
Caffeine	0.085	0.196	0.066	0.152
Carbohydrates:				
Glucose	0.30	0.69	0.10	0.23
Sucrose	Nil	Nil	Nil	Nil
Starch	6.10	14.09	6.14	14.22
Pectins	2.25	5.20	4.11	9.52
Fibre	2.09	4.83	2.13	4.93
Cellulose	1.92	4.43	1.90	4.39
Pentosans	1.27	2.93	1.21	2.80
Mucilage and Gums	0.38	0.88	1.84	4.26
Tannins:				
Tannic acid	2.24	5.17	1.99	4.61
Cacao-purple and cacao-brown	5.30	12.26	4.16	9.63
Acids:				
Acetic (free)	0.014	0.032	0.136	0.315
Oxalic	0.29	0.29	0.30	0.70
Extracts:				
Cold water	15.28	35.28	12.29	28.46
Alcohol (85%)	16.80	38.80	9.67	22.39

Chocolate manufacturers convert cacao beans into chocolate, and various chocolate-based products. Further, the chocolate melts at the ambient temperature of the cocoa bean producing countries. As a result, the bulk of the consumption of cocoa products is in the non-producing temperate and subtropical countries. Consequently, manufacturers in the final analysis have used the simple law of supply and demand to hold, in many cases, the cocoa bean producers to economic ransom. They have developed a chain of cocoa forecasters who go round the major cocoa producing countries of the world each year to forecast the crop. The sponsors of these forecasters (the buyers) base their buying policies on these crop forecasts.

7.9.2 The Cocoa Bean Testa

The cocoa bean testa (shell) is an important by-product in the chocolate industry. Its composition is shown in Table 12.

The chemical composition of the shell shows that it can be hydrolysed under pressure with sulphuric acid to produce fermentable sugars, but this may not be economical. With its low fibre content, freedom from alkaloids and good ash component, the shell is a suitable ruminant feed.

The shell is a good source of vitamin D. It can also be used as organic fertilizer. It has been so used in the United States of America and Great Britain to lighten heavy soils.

Physically the shell can be used for board material, but it lacks the tensile strength and resilience needed for durability.

The shell has been shown to have a calorific value ranging from 16,000 to 19,000 BTU per kilogram and this is a little higher than the value for wood.

7.9.3 The Rejects and the Pod Husks

One of the major problems facing cacao products is the economic use for various by-products of cocoa, hitherto completely wasted. These by-products are the cocoa pod husk (56 per cent of the pod), the sweatens (mucilage) (2.00 per cent) and unfermentable black pod disease infected cacao pods.

The black pod disease infected cacao pods have been analysed and found virtually not to differ chemically from normal mature cacao pods. Unfermentable beans are not acceptable to the world market because the chocolate aroma cannot be easily imparted to them. The chocolate aroma of cocoa beans develops during fermentation. Mucilage of unfermentable beans is scanty and contains less of the simple sugars. The pod husks in both cases are chemically the same.

Table 12 Chemical composition (%) of the cocoa testa (shell)

Water	3.8	Sucrose	0	Fibre	18.6
Fat	3.4	Glucose	0.1	Cellulose	13.7
Ash	8.1	Starch	2.8	Pentosans	7.1
Nitrogen	2.8	Pectin	8.0		

Table 13 Approximate chemical compositions (%) of (a) cocoa pod husk; (b) *Panicum maximum* (Guinea grass); and (c) *Centrosema pubescens* (a herbaceous legume)

	Cocoa pod husk (mature)	*Panicum maximum* (at 25 weeks)	*Centrosema pubescens* (at 25 weeks)
Water	57.75	63.40	69.40
Total dry matter	42.25	36.60	30.60
Crude protein	9.69	3.90	16.00
Pure protein	—	(3.40)	(14.70)
Fatty substances	0.15	0.80	1.80
Ash (SIO_2 free)	10.80*	3.60	6.20
Crude Fibre	*33.90*	*33.80*	*31.30*
Nitrogen-free extracts	42.21	48.80	43.00
Glucose	1.16	—	—
Sucrose	0.18	—	—
Pectin	5.30	—	—
Theobromine	0.20	—	—

*Contains the silica component

An approximate chemical composition of the cacao pod husk is recorded in Table 13. In comparison, approximate chemical composition of *Panicum maximum* (Guinea grass) and *Centrosema pubescens* (a herbaceous legume), both being well-known pasture components, are also recorded.

The cocoa pod husk has a low alkaloid content while tannin is practically absent. The crude fibre content is low; it is completely unlignified and compares favourably with *Panicum maximum* and *Centrosema pubescens*. It therefore appears that the pod husk should be digestible by all classes of livestock, especially the ruminants. In fact, Dittmar (1956) successfully fed cocoa pod husks to pigs (non-ruminant) in Brazil. The crude protein content of the cocoa pod husk is higher than that of the Guinea grass. The crude fibre of the Guinea grass and the *Centrosema* becomes progressively lignified with age, thus rendering them less digestible, whereas that of the cocoa pod husk is not lignified. The silica content is much lower than in Guinea grass.

The use of cocoa pod husk as a feed should be of special significance to West Africa in view of livestock development. Most cattle in West Africa are raised in the relatively dry north because tsetse fly has limited cattle rearing in the south. In the north, grazing is acutely short during the dry season which may last for four to six months of the year, while in the south over 1.5 million tonnes (dry weight) of cocoa pod husks are wasted annually! From the analytical data presented above, it appears that cocoa pod husk could be very nourishing (at least to the ruminants – cattle, sheep and goats particularly during the dry season when they subsist on more or less dry grasses). The collection, drying grinding, packaging and marketing of cocoa pod husk as livestock feed require urgent consideration.

Table 14 Composition (%) of the ash of cocoa pod husk

Constituent	Percentage (range)
CaO	0.22–0.59
MgO	0.40–0.52
K_2O	2.85–5.87
P_2O_5	0.30–0.49
SiO_2	0.06–0.14

The cocoa pod husk can be hydrolysed under pressure for fermentation into alcoholic drinks. This may not be economic because of other cheaper sources of alcohol, e.g. the waste products of the sugar industry.

The chemical composition of the ash from pod husks is given in Table 14. The silica (SiO_2) content is low. Potassium oxide (K_2O) is the only readily soluble oxide in the cocoa pod husk ash. Extraction of the K_2O in the form of potassium hydroxide (KOH) can be easily achieved by leaching the ash with water. The extracted KOH can be usefully employed in soap making.

Soap making is saponification of fat with a suitable hydroxide. Fat (cocoa butter fat) is obtainable from unfermentable cocoa beans which could easily be extracted from cocoa pods infected with black pod disease. This process has been investigated in detail at the Cocoa Research Institute of Nigeria and the results have proved satisfactory.

7.9.4 The Cocoa Sweatens (Mucilage)

The cocoa sweatens (mucilage) is a very interesting by-product of the cocoa industry (see Figure 26). It is a viscous liquid surrounding the seeds within the pod. On extraction of the seeds from the pod, the mucilage flows out. Sometimes, children collect the sweatens for drinking because it is very sweet (high sugar content). Apart from this, the sweatens has hitherto not been utilized. The chemical composition is given in Table 15.

The mucilage does not gel readily on heating. It is free of alkaloids and toxic substances. The two major components of the total dry matter (15.8 per cent) of the mucilage are pectin (5 per cent) and glucose (11 per cent).

Pectin is a simple relatively indigestible carbohydrate. It is a major component of jams, jellies and weight reducing dietary formulations. It is readily precipitated

Table 15 The chemical composition (%) of cocoa mucilage (sweatens)

Water	79.2– 84.2	Sucrose	0.11–0.90
Dry substance	15.8– 20.8	*Pectin*	*5.00–6.90*
Non-volatile acids	0.77– 1.52	Starch	— —
Volatile acids	0.02– 0.04	Proteins	0.42–0.50
Glucose	11.60–15.32	Ash	0.40–0.50

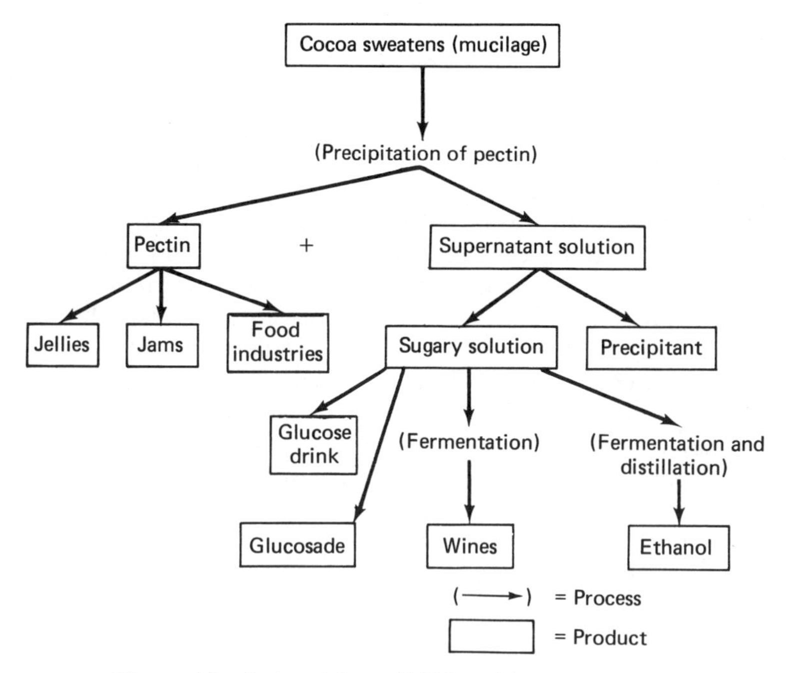

Figure 26. Industrial possibilities with cacao sweatens

from solutions and gels readily on heating. The pectin in cocoa mucilage can be precipitated by one of the customary and recoverable pectin precipitants. On recovering the precipitant, the remaining solution is composed of water, glucose and traces of sucrose, proteins and ash. The solution can therefore be sterilized, diluted to taste and bottled as a glucose drink or glucosade. The glucose in the solution can be crystallized out and be pharmaceutically employed.

On the other hand, the apparent glucose solution can be fermented to produce wines and other types of alcoholic drinks.

The only practical problem in the commercialization of this process is the collection of cocoa sweatens from farmers and adequate preservation of the material before processing. Studies have been carried out to find a reasonable solution to this practical problem. The sweatens should not be allowed to undergo fermentation before processing. Collection and storage of sweatens without fermentation is achieved by inoculating collecting cans with anti-fermentation preservatives. At the Cocoa Research Institute of Nigeria 1 g of sulphite to 5 litres of sweatens has been found satisfactory.

The future of cacao production in West Africa remains unpredictable. West Africa is undergoing rapid economic and industrial developments; suitable cacao soil is getting short; the population is increasing at a rapid rate, and this necessitates the use of more land for growing locally consumed food crops; other commodity crops are gaining international marked recognition as well as local importance. All these are competing with cacao for factors of production. Cacao will continue to be produced, but cacao cultivation in West Africa requires well designed and executed breeding; agronomic, disease and pest control; economic, environmental and utilization research; and the increasing application of proven results to production. So far, from the cacao pod, only the cacao beans are economically used. The cacao husk and the cacao sweatens have been shown to be capable of a number of industrial processes (see above). The physiologically mature but diseased cacao pods are currently thrown away. These pods contain physiologically mature cacao beans with high quality butter fat content. Loss to the world cacao production through these pods is about 25 to 30 per cent. Research and development of methods of retrieving the butter fat in these discarded diseased pods for industrial uses are required. The increasing cost and shortage of labour in West Africa call for mechanization of cacao production processes on the farm. The initial effort in the development of cacao cultivation in West Africa was taken, and has been sustained so far, by the peasant farmers. At this stage, increasing support for cacao cultivation by Government and international organizations concerned with cacao is needed if the crop is to survive long.

ADDITIONAL READING

Alvim, Paulo de T. (1972). *Factors affecting Flowering of the Cacao tree. Cocoa Growers' Bulletin*, Publication of Cadbury Brothers Ltd, Bournville, UK.

Annual Reports of the Cocoa Research Institute of Ghana (*CRIG*). Tafo, Ghana, 1963–1976.

Annual Reports of the Cocoa Research Institute of Nigeria (*CRIN*). Ibadan, Nigeria, 1963–1977.

Annual Reports of the West Africa Cocoa Research Institute (*WACRI*). Tafo, Ghana, 1950–1962.

Are L. A. (1967): Attempts to improve the performance of budded cacao materials in the green house: effects of rootstock, age, and time of budding. *Proc. 2nd International Cacao Research Conference*, Bahia, Brazil.

Cadbury Brothers Ltd (1963). *The Samoan Cacao Drier*. Cadbury, Bournville, UK.

Cocoa, Chocolete and confectionery Alliance (1959). *Raw Cacao*: *Manufacturers' quality requirements*.

Cocoa Growers' Bulletin (Quarterly Publication of Cadbury Brothers Limited, Bournville, UK).

Cocoa Research Institute of Nigeria: Advisory Leaflets Nos. 1, 2, 3, 4, 5, 6, 7, 9, and 11. CRIN, Ibadan, Nigeria.

Cocoa Research Institute of Nigeria: Memoranda No. 1 to 18. CRIN, Ibadan, Nigeria.

Dittmar, F. H. K. (1956). *The Value of Cocoa Husk in Pig Production*. Centro de Pesquisas do Cacau, Bahia, Brazil.

Erneholm, I. (1948). *Cocoa Production of South America*. Gothenburg, Holmquist.

Gibberd, A. V. (1951). Regeneration of cocoa farms in Nigeria. *Proc. West. African Cocoa Conference*, London.

Glendinning, D. R. (1965). Further observations on the relationship between growth and yield in cacao varieties. *Duphytica*, **15**, 116–127.

Gregory, P. H. (1974). *Phytophthora Disease of Cacao*. Longman, London.

Hall, T. H. R. (1963). The cutting production and rooting potential of some WACRI cocoa clones. *Tropical Agriculture*, **40**.

Hardy, F. (1960). *Cacao Manual.* Publication of the Inter American Institute of Agricultural Sciences, Turialba, Costa Rica.

Howat, G. R., Powell, B. D. and Wood, G. A. R. (1957). Experiments on Cacao Drying and Fermentation in West Africa. *Tropical Agriculture*, **34**, 249–259.

Knight, R. and Rogers, H. H. (1955). Incompatibility in *Theobroma cacao. Heredity* **9**, 69–77.

Longworth, J. F. (1963). The effect of swollen shoot disease on mature cacao in Nigeria. *Tropical Agriculture*, **40**, 275–283.

Longworth, J. F. and Thresh, J. M. (1963). The reaction of different cacao types to infection with swollen-shoot virus. *Annals of Applied Biology*, **52**, 117–124.

Opeke, L. K. (1969). Development of cacao industry in Nigeria. *Proc. Agriculture Society of Nigeria*, **6**.

Opeke, L. K. (1971). *Progress in Tree Crop Research in Nigeria.* Cocoa Research Institute of Nigeria, Ibadan, Nigeria.

Opeke, L. K. (1973). *Induced Mutations in Vegetatively Propagated Plants and Mutation Breeding in Cacao* (*Theobroma cacao L*). IAEA-PL-501/14, 137–143.

Opeke, L. K. and Gorenz, A. M. (1974). Phytophthora pod rot. Symptoms and economic importance. In *Phytophthora Diseases of Cacao* (ed. P. H. Gregory). Longmans, London.

Powell, B. D. (1958). The rapid artificial drying of cacao beans and chocolate flavour. *Tropical Agriculture*, **35**, 200–204.

Russell Cook, L. (1972). *Chocolate Production and Use.* Books for Industry Inc., New York.

Swarbrick, J. T. (1965). Storage of cocoa seeds. *Experimental Agriculture*, **1**, 201–207.

Selection of Soils for Cacao: *Soils Bulletin* (1966). FAO, Rome, Italy.

Thresh, J. M. (1958). *The Control of Cacao Swollen Shoot Disease in West Africa* (A review of the present situation). WACRI Tech. Bulletin No. 4.

Urquart, D. H. (1956). *Cacao.* Longmans, Green & Co., London, UK.

Wadsworth, R. V. (1955): The quality of raw cacao as it affects the manufacturer. *Tropical Agriculture*, **32**, pp. 1–9. Cacao soil Survey – CRIM, Ibadan, Nigeria.

Wessel, M. (1966). Review of the use of fertilizers on cacao is Western Nigeria *CRIN Annual Report 1965/66.*

Wessel, M. and Toxopeus, H. (1967). Seasonal influences on pod and bean values of West African Amelonada cacao. *Proc. 2nd International Cacao Research Conference*, Bahia, Brazil.

Wood, G. A. R. (1976). *Cocoa.* Longmans, London.

Wood, G. A. R. (1961). Experiments on cacao drying in the Cameroons. *Tropical Agriculture*, **38**, 1–11.

CHAPTER 8

*Kola**

*This chapter closely follows the monograph on kola by van Eijnatten (1969).

Kola, a member of the family Sterculiaceae, has a long history in West Africa. The use of kolanuts, the major product of kola, is intimately interwoven with the various cultures of the peoples of West Africa. The use of kolanuts features prominently in religious, social and ritual activities of West Africa. They are used during ceremonies related to marriage, child naming, funerals and in sacrifices made to the various gods and goddesses of African mythology.

8.1 ORIGIN OF KOLA

Johannus Leo Africanus was the first to refer to the kolanut in 1556 (Russell, 1955). The Portuguese Odoardo Lopez recorded the occurrence of kola trees in the Congo, in 1591, followed by Andre Alvares, who saw them in Gambia and Guinea in 1594 (Nzekwu, 1961). Subsequently, the tree was recorded all along the west coast of Africa from Gambia to Angola. The early records did not distinguish between the two commercial species of kola – *Cola nitida* (L.) and *Cola acuminata* (L.).

C. nitida was originally distributed along the west coast of Africa from Sierra Leone to the Republic of Benin (Bontekoe, 1950; Chevalier et Perrot, 1911; Nzekwu, 1961) with the highest frequency and variability in the forest area of Ivory Coast and Ghana. This area has now been accepted as the centre of origin of *C. nitida*. This same area remained for long the only source of kolanuts (gbanja kolanuts) to the West African trade routes. In the early twentieth century, kola trees were seldom planted, the nuts used for trade and local consumption being obtained from spontaneously occurring trees (Miles, 1930).

C. acuminata, another important commercial species of kola, has its original area of distribution stretching from Nigeria to Gabon. Conde de Ficalho (1945) stated that *C. acuminata* occurred spontaneously in mountainous areas of Angola, Zaire and Cameroun, while it has long been in cultivation on the islands of Principe and Sao Thome. Southern Nigeria is currently regarded as the centre of origin of *C. acuminata*.

On account of the importance of kolanuts in the early trade of West Africa, the two important commercial species of cola were redistributed from their centres of origin through the activities of man. Chevalier et Perrot (1911) and Warburg (1902) stated that *C. nitida* was carried through Nigeria towards Cameroun and Congo around 1900. By 1908, the cultivation of *C. nitida* had been firmly established in Nigeria. Cultivation of *C. nitida* also spread westwards reaching its limit in Gambia and Guinea, By the middle of the twentieth century, the cultivation of *C. nitida* had spread from its centre of origin westwards to the southern border of Senegal with Gambia, eastwards into Zaire and also, overseas, to the Caribbean islands (especially Jamaica). *C. acuminata* also spread to other parts of West Africa. In 1900, it was introduced into Ivory Coast.

The estimated production of kolanuts in the major producing areas (1966 and 1976) is shown in Table 16. It must be noted that production increased by almost 24 per cent between 1966 and 1976.

Table 16 Estimated production of kolanuts in 1966 and 1976 (tonnes wet weight)

	1966	1976
Ivory coast	14,500	17,000
Guinea	4,342	5,500
Cameroun, Benin, Togo	11,000	13,000
Ghana	14,000	16,000
Sierra Leone	4,000	5,000
Nigeria	120,000	150,000
Liberia	4,000	5,000
South America	3,000	4,500
TOTAL	174,842	216,000

8.2 TAXONOMY

Several attempts have been made to describe correctly the taxonomic position of the genus *Cola*. Prominent among the botanists who have tackled this problem were Schumann (1900), Chevalier and Perrot (1911), and Bodard (1962).

8.2.1 The Genus *Cola*

The family of the Sterculiaceae was created by Ventenat (1804) within the Order of the **Malvales**. The *Sterculeae* forms a tribe (with unisexual flowers without petals) within the family. The genus *Cola* is the most important within the tribe *Sterculeae* in Africa. Schott and Endlicher established the genus *Cola* in 1932. Before then, a few *Cola* species has been known under the generic name *Sterculia*. During the nineteenth century, many *Cola* species were discovered and described. About 50 *Cola* species have now been recorded and identified in West Africa (Onochie and Stansfield, 1960). Two species that had in the older days been grouped under *Cola* have in the modern classification been upgraded into two independent genera. These are *Ingonia* and the *Chlamydocola*.
The currently accepted definition of the genus *Cola* is as follows:

Trees and shrubs with alternate leaves; stipules present although sooner or later dropping. Male and hermaphrodite flowers grouped into a panicle of cymes, or in fascicles on the branches or on the trunks. Five sepals; male flower: the anther loculi are placed laterally at the top of the androecium in one or two superimposed rings; hermaphrodite flowers: five to ten carpels narrowly placed together in the centre of the flower, style short, stigma fleshy, more or less recurved, a vestigial androecium at the base of the gynaecium. Fruit: five to ten follicles, placed perpendicularly on the peduncle, dehiscent at maturity, containing one to twelve seeds; seeds fleshy, without endosperm, radicles directed toward the hilum. After germination of the seedling the cotyledons are subhypogynous.

Bodard (1962) recognizes five subgenera, of which the subgenus *Cola Bodard* (= Eucola A. Chevalier) includes the species with edible nuts. The most important of these are *C. nitida* (Ventenat) Schott and Endlicher, and *C. acuminata* (P. Beauv) Schott and Endlicher. The first one is the major economic crop, but *C. acuminata* is of some importance in Southern Nigeria. The remaining species in the subgenus *Cola* are not cultivated, but their nuts are sometimes used to adulterate the commercial produce in times of scarcity. This is particularly true for the species *C. ballayi* and *C. verticillata*.

8.2.2 *Cola Nitida* (Ventenat) Schott and Endlicher

Various synonyms are listed by Hutchinson and Dalziel (1958).

1. *Sterculia nitida* Ventenat
2. *Cola vera* Schumann
3. *Cola acuminata* (Beauvour) Schott and Endl. var. *latifolia* Schumann
4. *Cola acuminata* Engler.

Russell (1955) published the following description of *C. nitida*:

The Gbanja kola tree is a robust tree, usually from 9 to 12 m high, although sometimes attaining a height of 24 m. The tree may have narrow buttresses extending up the trunk for less than 1 m. The foliage is dense and not confined to the tips of the branches. The leaves have long petioles with prominent pulvinae and are rather variable in size and shape, commonly obovate, sometimes rather long, with the base of the lamina usually cuneate, the tip abruptly acuminate. The leaf is flat except that the tip is sometimes slightly bent; the lateral veins are very distinct, making furrows in the upper surface of the leaf.

The inflorescence is non-verticillate, the flowers rotate or slightly cupped. Hermaphrodite flowers are commonly 3 cm long, and may be up to 5 cm across; male flowers are smaller, up to 2 cm across. In both cases the perianth is segmented for at least two-thirds of its width. The perianth is white or cream coloured with a small dark-red marking at the base of each segment inside, extending a short distance up the three veins; it is glabrous inside or with only a few scattered stellate hairs. There is little or no column to the androecium, which is almost sessile at the base of the perianth and is more or less red in colour.

The fruit is composed of five follicles borne on a short pendant peduncle, the follicles usually horizontal or in a recurved position. The follicle is roughly ovoid but curved; the more convex ventral (adaxial) side has a pronounced keel, which is produced into a short, terminal curved beak. The follicles usually dehisce when mature, along the ventral suture. The follicles are up to 13 cm long and up to 7 cm wide; when nearly mature they are shining green, though sometimes nearly white when young, and the surface is often rugose or tuberculate. Each fruiting carpel contains up to ten seeds in two rows. The embryo is covered with a tough membranous white or pink seed coat and the remainder of the nucellus tissue; it has two, occasionally three, cotyledons of a white, pink or red colour. The seeds are variable in size up to 5 cm long, ellipsoid, globular or angular by compression. The embryo constitutes the kolanuts. The kolanuts of this species usually mature during the months of November and December.

The occurrence of various frequencies of differently coloured nuts in a number of geographical areas of West Africa, caused Chevalier et Perrot (1911) to raise varieties or types of *C. nitida* producing variously coloured nuts to the status of subspecies. The validity of this was questioned by Bodard (1962). Apparently, no sharp distinction exists between the various purported subspecies, neither in nut colour, nor in geographical distribution. This character on its own is unlikely to be a sound basis for recognizing subspecies or even botanical varieties, because the relative frequencies of various genes controlling the nut colour, are subject to sudden changes by incidental or planned selections.

For the recognition of differentiation within the species more detailed observations will have to be made on the types of *C. nitida* occurring in various areas. The combination of differentiation characteristics may lead to an understanding of the evolution of the species. One such characteristic was recorded by Bodard (1962) who noticed that *C. nitida* nuts from Ivory Coast open up the cotyledons when germinating, while Dahomeyan nuts do not, thus causing the seedlings to emerge from the base of the nuts. This is, in fact, also the case with Nigerian and Sierra Leonean kolanuts. Other characteristics are the depth of the basal furrow in the cotyledons, which differ distinctly for three types of kola in

Nigeria (van Eijnatten, 1966), and the rapidity of germination of freshly harvested kolanuts (van Eijnatten, 1967).

8.2.3 *Cola acuminata* (Beauvoir) Schott and Endlicher

Hutchinson and Dalziel (1958) give two synonyms:

Sterculia acuminata Beauvoir.

Cola pseudoacuminata Engler.

Russell (1955) also published a description of *Cola acuminata*. As this tree is very similar to *C. nitida*, the differentiating characters only are presented here: The abata kola tree (*C. acuminata*) is a slender tree, up to 12 m high, but usually 6 to 9 m. The branches are slender, crooked and markedly ascending; the foliage is often sparse and confined to the lips of the branches.

The hermaphrodite flower may be up to 25 cm across; the male flower less. The perianth segments are usually joined for nearly half their length. The anthers are borne on a short, but distinct column.

The fruit consists of five follicles borne at right angles to the stalk or slightly bent downwards. The follicles are sessile and have a straight point or tip, the whole being up to 20 cm long. The surface is rough to the touch, russet or olive-brown. There are up to 14 seeds in each follicle. The embryo – the abata kolanut – may have three to five or even six cotyledons, pink, red or sometimes white in colour.

The fruits mature in the period from April to June.

8.3 THE FLOWERING PROCESS

Both *C. nitida* and *C. acuminada* possess male and hermaphrodite flowers. The flowers are arranged in determinate, paniculate cymes. The male flowers possess a rudimentary gynaecium which is non-functional while the hermaphrodite flowers possess a developed androecium which does not function and a functional gynaecium. The two types of flowers occur mixed in the same inflorescence, but occasionally they may occur in different inflorescences or trees.

8.3.1 Time of Flowering

The main flowering season of kola lasts approximately three months. Within this period 80 to 90 per cent of the total number of flowers produced, emerges. In Nigeria this occurs from August to October with a peak in September.

In the Central African Republic the flowering season is from July to September (Dublin, 1961). The proportions of male and hermaphrodite flowers are influenced by the time of flowering. In Nigeria, 12 per cent of observed flowers are hermaphrodite. During the flowering period this proportion varies from 11 to 13 per cent. However, during the period from January to June, when only few inflorescences are produced, this proportion proves to be much higher: from 24 to 83 per cent. Dublin (1961) observed that in the Central African Republic the proportion of hermaphrodite flowers decreases during the flowering period, from 28 per cent in July to 12 per cent in September.

Individual trees vary greatly in the percentage of flowers produced during the flowering period; few trees carry up to half their flowers outside this period. Also the proportion of male and hermaphrodite flowers is variable. Some trees produced only male flowers, others 50 per cent or more of hermaphrodite flowers (Chevalier and Perrot, 1911; Russell, 1955).

8.3.2 Differentiation of the Flowers

Some weeks prior to the shedding of the first bract, the inflorescence bud becomes thickly rounded. Shedding of the first or outer bract of this bud marks the first day that the inflorescence starts its development. Prior to this moment the bud has a globular primordium, which develops into an inflorescence with partly differentiated flowers in the period between shedding of the first, brown and coriaceous bract and the second, white bract. This period lasts about three days.

When the second bract is shed, the inflorescence becomes visible and the flowers continue their differentiation. The sepals are formed first, followed by the androecium and later the gynaecium.

The reduction division in the pollen mother cells of the most advanced terminal flower coincides with shedding of the second bract surrounding the inflorescence. During the first metaphase of the reduction division 20 bivalents become visible: the somatic chromosome number ($2n$). Four to five days after the reduction divisions of the pollen mother cells the pollen nuclei divide into two. It is only at this time that the gynaecium starts to develop up to the formation of the ovules. Subsequently, ovules are formed in the hermaphrodite flower, while in the male flower the development halts at this stage and, therefore, its gynaecium consists of sterile carpidia (see Figure 27).

When the flowers are ready to open, the male flowers have a subspherical shape and measure from 12 to 20 mm in diameter. The hermaphrodite flowers have at this stage more oval shape and are larger: 30 to 40 mm in diameter (Bodard, 1962; Russell, 1955). The flowers have five connate sepals, each with three parallel red lines, which vary in width and length. Occasionally this red coloration may be absent altogether, while also at the other extreme, complete red coloration of the adaxial side of the sepals has been observed. Petals are absent. The androecium consists of five connate groups of four pollen sacs, alternating with the five staminal groups, which sporulate freely.

In the hermaphrodite flowers the pollen sacs are indehiscent and do not sporulate. The five carpels of the hermaphrodite flower contain 10 to 12 anatropous ovules, implanted along the ventral, adaxial side of the carpels.

The shape and size of both male and hermaphrodite flowers are subject to large variation. In addition to this, many aberrant types of flowers have been

Figure 27. Floral structures of kola

recorded: various grades of development in the gynaecium of the male flowers, irregularly placed stamens, fusion of superimposed pollen sacs, presence of petals, virescent flowers, multiple flowers (Bodard, 1955, 1962). These abnormal types of flowers may occur on any kola tree.

8.3.3 Pollination

8.3.3.1 Anthesis

Anthesis occurs early in the morning from 4.00 to 5.00 a.m., up to around 8.00 a.m. The flowers start drying up at the end of the second, third or fourth day, although some flowers may persist for up to 10 days (Bodard, 1955, 1962; Paula, 1938). The number of hermaphrodite flowers, which actually set, is very low. Estimates of the numbers of hermaphrodite flowers developing into fruits were variously given as 1.6 per cent in the Central African Republic and 2.7 per cent in Ivory Coast (Bodard, 1962; Dublin, 1965).

The receptivity of the gynaecium decreases with time as shown by the percentages of fruits set in hermaphrodite flowers, hand pollinated at various periods from opening of the flower. On the day of anthesis 23 per cent of the pollinations proved successful. Those carried out on the fourth day had a success rate of 14 per cent. Later pollinations all failed. As we shall see later, various kola trees differ greatly in their capacity of serving as female parents or as pollinator parents because of various degrees of compatibility.

The anthers of male flowers are sporulating at anthesis and some of the pollen grains are deposited on the adaxial side of the sepals, just below the pollen sacs. The sticky pollen grains are quite large: 30.9 ($\pm$ 1.9) μm in diameter. The pollen sacs in hermaphrodite flowers do not sporulate; no pollen grains are liberated. The pollen grain from hermaphrodite flowers is elongated (prolate). Bodard (1962) studied the germination of both 'male' and 'hermaphrodite' pollen grains on a 1 per cent agar agar solution of 10 per cent saccharose. Both types of pollen grains germinated rapidly, within three hours. The viability of the pollen grains was about 80 per cent on the day of anthesis. It dropped to a level of 10 to 40 per cent during the second day and of 2 to 5 per cent after three days. On the seventh day no pollen grains were viable any more. Subsequently Bodard (1955) carried out hand pollinations and obtained fruits setting when 'male' pollen grains were used. No success ever followed if pollen from hermaphrodite flowers was used for either self-pollination or cross-pollination with other trees. Unless in the unlikely event that the trees used by Bodard for this investigation were all cross-incompatible, these facts indicate that pollen grains from hermaphrodite flowers are not functional, although viable. Therefore, hermaphrodite flowers can be considered, functionally, as female flowers.

8.3.3.2 Conveyance of pollen grains

Because of the sticky nature and the size of the pollen grains, natural pollination is unlikely to be anemophilous. The flowers of kola have, in addition, a penetrating smell and both Russell (1955) and Bodard (1962) contend that flies of different kinds are attracted by this. These would then carry pollen from one tree to another. No proof has yet become available to substantiate this suggestion.

The mirid *Torma colae* China has, at one stage, been mentioned as the elusive pollinator of the kola flower, but Bodard (1962) considered this unlikely. *Torma colae* remains for the major part of its life cycle within the kola flower and is unlikely to carry pollen grains over long distances. That this may occur was shown by Voelcker (1935), who observed red-coloured kolanuts in fruits from a tree raised from a white nut homozygous for the recessive white nut colour factor while the closest possible pollinator parent raised from a coloured nut was growing at a distance of 270 m.

The low percentage of fruit setting resulting from natural pollination has as yet prevented conclusive information to be obtained on some characteristics of the pollinator, which apparently is active in the daytime. Flowers covered with a paper bag from 6.00 a.m. to 6.00 p.m. never set any fruits, while flowers bagged from 6.00 p.m. to 6.00 a.m. did set fruits occasionally. Similarly, it is apparent that the pollinator is winged, as fruits developed on inflorescences which have been kept ringed with bands of grease to prevent access to the flowers by crawling insects. Workers at Ibadan have suggested that a *Forcipomya* sp. may be responsible for the pollination of kola flowers, as in cacao.

The low efficiency of the pollinating agent in effecting pollination led Bodard (1962) to study the possibility of increasing the number of flowers and prolonging their life period. This would then increase their probability of being pollinated. The application of indole acetic acid, indole butyric acid or naphthalene acetic acid to individual inflorescence buds at the moment of the shedding of the bracts caused a rapid development of the inflorescence. However, the inflorescence remained compact and the flowers soon turned brown. The application of the same compounds as a dust or a spray in low concentrations (0.0010, 0.0025 and 0.0050 per cent) to the whole tree resulted in a much longer retention of the flowers, which remained viable for 15 to 20 days. Also the growth hormones 2, 4-dichloro phenoxy acetic acid and 2, 4, 5-trichloro phenoxy acetic acid had this effect when applied as a dust or a spray.

The effect of the longer viability of the flowers, especially on the receptivity of the hermaphrodite flowers and, therefore, ultimately on the productivity of the trees has not been studied as yet. Increasing the efficiency of natural pollination would provide means to improve the productivity of mature trees without recourse to controlled pollination, which becomes increasingly difficult with the height at which flowers are borne.

8.3.3.3 Controlled pollination

Controlled pollination has been suggested as a means of increasing the productivity of the kola tree. This technique is used extensively for experimental purposes and some information has become available as to its efficiency.

The stigmas of the hermaphrodite flowers are most receptive on the day of anthesis and pollen grains have their highest viability on that day. Therefore, it is important to use hermaphrodite and male flowers on the day that they have opened. If, for experimental purposes, foreign pollen is to be excluded, the hermaphrodite flowers to be pollinated should be covered with a muslin or paper bag on the evening prior to the day of anthesis. Nearness of flower opening can be judged from the tension in the sepal furrows. The exterior edges of the sepals are separated with the inner ones still connate, when the flower is approaching anthesis.

When the hermaphrodite flower has opened, viable pollen grains should be obtained from male flowers to be applied to each of the five stigmas of the hermaphrodite flower as otherwise only a partially developed fruit will result. Bodard (1962) transferred pollen with the aid of camel hair brushes, dissecting needles or with the male flower itself. He found differences in fruit setting in favour of pollination with the aid of a dissecting needle. Similar experiments at Ibadan, Nigeria, did not confirm the existence of differences due to these methods of transferring pollen grains.

Hand pollination often raises the percentage of flowers setting fruit, considerably above the level attained by natural pollination (Bodard, 1962; Dublin, 1961; Paula, 1938; Russell, 1955). Studies at Ibadan, Nigeria, have shown that 3546 controlled pollinations within various cultivars of kola (Agege Red, KD Pink) led to a success in pollination varying from 24.6 to 34.5 per cent, while 2679 controlled pollinations carried out within the cultivar 'Gambari White' gave an average success of only 3.5 per cent. Apparently, large differences exist in the success in pollination to be expected within various populations of kola trees. Furthermore, various parents apparently differ in their ability to contribute to the success in controlled pollinations. The female parents proved to be of more importance in determining the success percentage than the pollinator parents, although, in particular combinations, various degrees of cross-compatibility were observed (van Eijnatten, 1967).

8.3.3.4 Incompatibility

Fertilization of hermaphrodite flowers with pollen from male flowers on the same tree sometimes results in fruit setting. Alternatively, cross-pollination among any two seedling trees often meets with success.

Self-incompatibility of kola is reported to be the rule in Ivory Coast, where Bodard (1962) obtained success in self-pollination only in the case of 2 out of 50 trees tested, while cross-pollinations showed a high degree of success. Although Russell (1955) stated that self-pollination usually is successful in Agege Red trees, subsequent observations have shown that this statement is to be reversed. Self-incompatibility occurs frequently also in various types of kola studied in Nigeria, 76 per cent of the trees being self-incompatible. Bodard (1955) found that pollination with compatible pollen a few hours after pollination with incompatible pollen leads to successful fruit formation. The incompatible pollen does not render the gynaecium of the self-pollinated flower incapable of receiving compatible pollen.

Experimental cross-pollinations between selected kola trees have shown that many trees are not compatible. This information became available from a systematic pollination programme at Ibadan, Nigeria, in which 53 selected kola trees were to be crossed in all possible combinations. The overall success percentage in pollination was 18 per cent. Exclusion of all incompatible combinations raises the average success to 32 per cent. The high number of incompatible combinations is remarkable. Only 25 of the 253 incompatible combinations observed refer to self-pollinated trees. The remaining ones prove that cross-incompatibility is also a serious problem in kola with regard to productivity.

Finally, the degree of compatibility varies considerably among parents. This can be illustrated by the results of an experiment in which 19 selected Agege Red kola trees were used as female parents to be fertilized with pollen from three other trees (van Eijnatten, 1966b). On average 14 controlled pollination were carried out per combination. Analysis of the data showed that there were differences between the female parents ($p < 0.001$), but that the male parents gave similar successes in pollination (ranging from 28.0 to 40.5 per cent, with an average of about 34.5 per cent). The degrees of compatibility of the female parents with the three pollinator parents used in the programme, as expressed in the success percentages in pollination, ranged from 10 to 90 per cent.

Although in other pollination programmes differences between male parents may be recognized as significant, usually the differences between female parents are responsible for a much larger part of the variance than those between the males.

Incompatibility, apparently, plays an important role in kola, both within and between trees. The compatibility within and between selected trees has to be studied, especially if the use of clones for the economic production of kolanuts is to be adopted.

8.3.4 The Development of the Fruit

Taking the day that the outer bracts of the inflorescence buds are shed as the first day, fertilization of the terminal flower should occur on the thirteenth to fifteenth day. One or two weeks later the carpels will change their upright position to a lateral one, thus already showing the star-shaped outward appearance of the full-grown kola fruit. The young fruits may point in any direction depending on the location of the inflorescence. While the fruit is increasing in weight, the carpels turn downwards with the adaxial side of the carpels below.

A heavy loss of fruits in the period from two months after pollination to maturity was reported by Russell (1955), who blamed insect pests for this phenomenon. Later observations in Nigeria have shown that the majority of the young fruits are lost earlier.

In 1965 artificial pollination resulted in 734 apparently developing fruits at the end of the second week after pollination. During the course of the next 8 weeks, from the second to the tenth week after pollination, 193 fruits or 26 per cent were lost. A small part of these had been damaged by caterpillars eating through the little pods. Most fruits, however, dropped without apparent reason (see Table 17).

Before six weeks had passed after pollination, already 71 per cent of the dropping fruits had been lost. The remainder dropped during the following four weeks. From the tenth week onwards no further fruits were lost.

As Russell (1955) states: 'In kola the number of fruits set is normally small in relation to the bulk of the tree, and the water supply is abundant towards the end of the rainy season as the fruits are developing, so that one would hardly expect

Table 17 Fruit drop at various times after pollination

Time after pollination (weeks)	Fruits dropped	
	Number	Percentage
2 to 4	91	47
4 to 6	46	24
6 to 8	36	19
8 to 10	20	10
Total	193	100

normal fruit production to impose undue strain on the resources of the trees.'

However, the location of the fruits within the tree has proven to be a major determinant of the size of the nuts and, therefore, also of the size of the fruit. These considerations lead to the supposition that indeed there are physiological limits to the development of the fruit, but that they are determined by much more localized factors than, as Russell puts it, the bulk of the tree. It may be that the retention of fruits and the size of retained fruits depend on the leaf area of the branchlet from which the fruit is developing.

The fruit developing from the five carpels, consists of five follicles, if the stigmas were all successfully pollinated. The follicles have an adaxial or ventral suture, at both sides of which four to six seeds are arranged alternately. The implantation of the funiculi of the seeds is clearly distinguishable. The dorsal or adaxial follicle wall contains one main vascular bundle, while the ventral side contains two well-developed vascular bundles feeding the two placentae, which develop on the margins of the carpels. The dorsal side of the follicle may be smooth, but is usually rugose. The degree of knobbiness and the size and shape of the follicles are characteristic for particular trees and varies greatly from tree to tree.

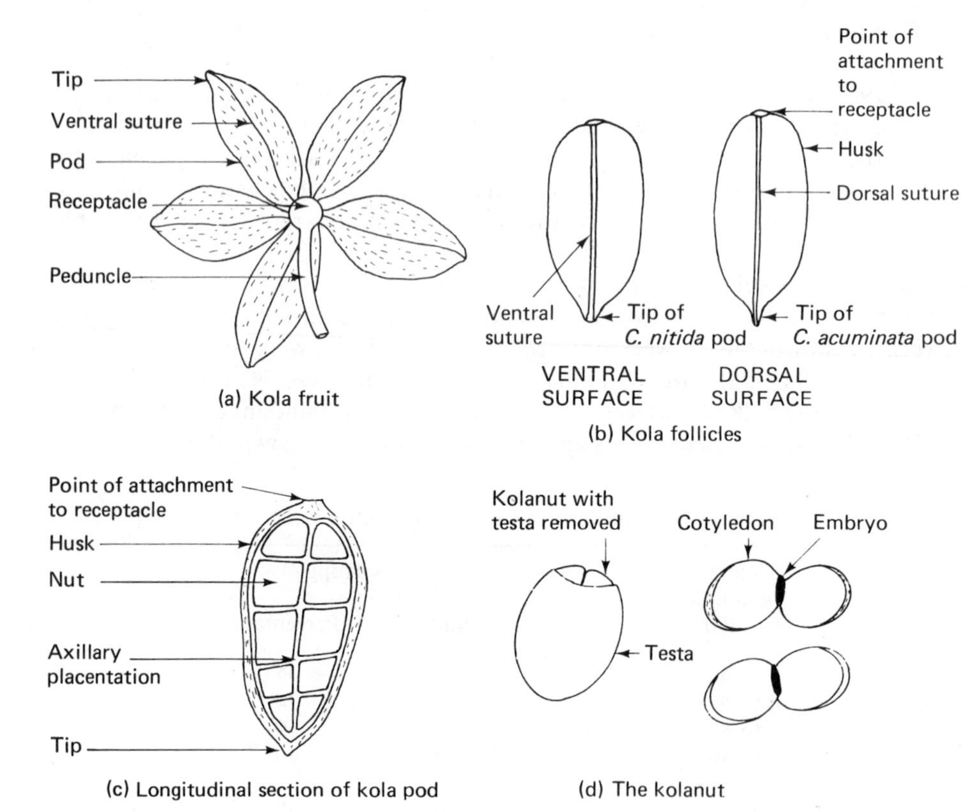

Figure 28. Diagrams of kola fruit, follicle, seed and kolanut

Maturing of the fruit is reached after 120 to 135 days (Bodard, 1962; Paula, 1938), although maximum weight is already achieved somewhat earlier. At maturity the follicles usually dehisce along the ventral suture, splitting between the two well-developed ventral vascular bundles. Sometimes fruits may dehisce prior to maturity; this occurs irregularly and the split occurs at the point of highest tension independent from the ventral suture. Some trees have apparently indehiscent fruits (Dublin, 1961).

Follicles may vary greatly in weight, depending on various factors among which are the vigour of the tree on which they grow, its genotype as well as that of the pollen parent, the fertility of the soil and water relations.

Dublin (1961) weighed a large quantity of follicles and found an average weight of 262 g with a range from 25 to 850 g. The follicle wall accounted on average for 42 per cent of the follicle weight. These figures explain the necessity of the rapid expansion of the brittle rachis of the inflorescence to a thick woody fruit stalk which may reach a diameter of 12 to 15 mm. Also the young wood on which the fruits develop is thus fortified.

8.3.5 The Development of the Seed (nut)

When the young seeds are developing in the follicles of the kola fruit, they have a very thick conspicuous seed coat and the embryos are located in the nucellus, a small, soft, opaque tissue. Approximately 80 days after pollination the embryos have assumed their final shape, colour and texture, although they are still very small. The remainder of the nucellus can be found as a very thin membrane at the inner side of the seed coat.

A study was made of the viability of kolanuts (cultivar: Gambari White) during their process of maturation. Hand pollinations were carried out and inflorescences with more than one successful pollination, therefore resulting in multiple fruits with many follicles, were chosen. Individual follicles were removed at varying periods from pollination. The seeds were extracted from these follicles and planted. No germination occurred in kolanuts planted out at 63 days from pollination (Table 18). After 84 days 60 per cent had germinated and 20 days later 94 per cent had germinated. At this time the kolanuts apparently were not fully mature, as shown by the number of days taken to reach germination. The early harvested kola nuts (at 84 and 105 days from pollination) took 87 to 94 days to reach germination: this period became increasingly shorter

Table 18 Percentage germination and number of days to germination for nuts at varying stages of maturity (cultivar: Gambari White)

	Number of days from pollination					
	63	84	105	119	126	133
Germination percentage	0	60	94	94	96	100
Days from planting to germination	—	94(± 3)	87(± 5)	57(± 3)	31(± 1)	22(± 1)

until all kolanuts harvested at 133 days from pollination germinated within 22 days from planting. Full maturity of these fruits was therefore, not reached before 133 days from pollination.

The mature seeds are surrounded by a thick fleshy seed coat, which soon decays when exposed. In it is contained the embryo, denoted as the kolanut, which consists of two very large fleshy cotyledons united through a tiny plantlet, which is 0.5 to 1.5 mm long. This plantlet is usually indicated as the embryo, although in fact it only constitutes a part of it.

There is a large range in nut weight. Within one follicle the nuts at the extreme ends are usually smaller than those in the middle. The largest variation occurs, however, among nuts from different fruits, even within produce from one tree. Russell (1955) weighed many thousands of nuts harvested from kola trees at Agege, Nigeria, and estimated the average nut size at 16 g. Nuts weighing less than 1 g or up to 100 g may be obtained, but the more usual range lies between 5 and 30 g. (Bodard, 1962; Chevalier et Perrot, 1911; Paula, 1938; Russell, 1955).

8.3.5.1 Anatomy of the embryo

Freshly harvested and stored nuts show anatomical differences. Freshly harvested nuts (cultivar: Agege Red) have an embryo which is 1.5 times as long as broad with approximately six differentiated leaf initials. In the axils of the leaves pockets of meristematic tissue are left behind by the advancing vegetation point, being separated from the axillary meristems' undifferentiated rounded parenchyma cells.

The apical meristem of the radicle extends upwards into a cylindrical zone of three to four layers of long narrow procambium cells reaching into the hypocotyl. Some protoxylem vessels are present at the adaxial side of the procambium, leading into the cotyledons. The radicles have a small schizogeneous mucilage duct in the centre.

In stored nuts of the same cultivar the embryo is twice as long as broad and has up to 12 differentiated leaf initials, twice as many as in the fresh nuts. In the stored nuts all meristematic regions have become connected through a narrow zone of elongated procambium cells, with pronounced nuclei. In the radicle differentiation into xylem elements has started. The central core of the pith in both stem and radicle of the embryo, has developed into long narrow cells situated horizontally. In various places these cells were swollen and had ruptured walls, thus giving rise to a long, central, schizogeneous mucilage duct.

These changes may be related to the apparently obligatory dormancy period, which kolanuts need prior to their being able to germinate rapidly.

8.3.5.2 Colour of the nuts

Variously coloured nuts (red, pink and white) may occur on one tree and even within one fruit. This depends on the genotype of the nuts which constitute the new generation. The inheritance of nut colour has been insufficiently studied.

Simple segregation ratios have been found in a few instances (Voelcker, 1935) which showed that red and pink are dominant to white. Van Eijnatten (1965) records a segregation of nuts into 3 red to 1 pink in the self-pollinated tree AD44 (at Agege, Nigeria). Self-pollination of trees established from white nuts usually results in white-nutted progenies (Voelcker, 1935). Tree AD44 therefore seems to be heterozygous for a red nut factor, the double recessive being white. White nut colour is probably recessive to pink. Another tree KD21 (Moor Plantation, Ibadan) segregated into 108 pink to 46 red nuts, when cross-pollinated with a white nut tree (double recessive); this shows that in this case pink nut colour was dominant to red, white being recessive to both pink and red. Many more involved ratios have been observed, where segregation cannot be explained by the assumption of two or three genes controlling the red or pink nut colour.

8.3.5.3 Chemical composition of the kolanut

Only approximately one third of the weight of the freshly harvested nut is accounted for by dry matter. The percentage of dry matter drops some 5 per cent during the first few weeks after the harvest and then gradually rises, depending on the conditions of storage (van Eijnatten, 1966). Dried kolanuts have been analysed on various occasions. The analyses reported in literature are usually fragmentary. Two analyses respectively reported in 1911 and 1953, gave information on the most important constituents, although 16 per cent of the dry matter was left unaccounted for (see Table 19).

Interest in the physiologically active agents led to the identification of alkaloids.

It is considered that in fresh kolanuts an unstable complex occurs as kolatin, a tannin, and caffeine glycosides. This complex oxidizes and hydrolyses to form kola red and free caffein under the influence of enzymes, when the nuts are drying out. If these enzymes are inactivated prior to drying the seeds, for instance with heat treatment, then this process does not occur and the dried seeds are said to retain their physiological action. Others state that the caffeine

Table 19 Chemical analyses of kolanuts; contents expressed in percentage of dry matter

	Chevalier et Perrot (1911)	Kolanuts and kola chocolate (1953)
Cellulose (fibre)	9.0	8.1
Carbohydrates	53.1	52.4
Fats	1.6	1.6
N. containing substances	10.8	11.1
Tannins	3.9	4.4
Ash	3.5	3.4
Caffeine	2.5	3.2
TOTAL	84.4	84.2

Table 20 Mineral content of kolanuts in percentages of dry matter (Sprecher von Bernegg, 1934; Egbe, 1966)

	Congo	Guinea	Nigeria
N	1.31	2.09	1.32
P	0.15	0.20	0.10
K	0.92	1.47	1.01
Ca	0.09	0.08	0.07
Mg	0.20	0.27	0.21

occurs partly free and partly in the above-mentioned complex (Boelman, 1940; Chevalier and Perrot, 1911; Kolanuts and kola chocolate, 1953; Kolanuts, 1959; Mascre and Paris, 1946).

The caffeine content of kolanuts ranges from 1.5 to 3.2 per cent of the dry matter. The usual range is from 1.5 to 2.7 per cent of the dry matter the value of 3.2 per cent being exceptionally high.

In addition kolanuts contain very small quantities of the alkaloids theobromine (0.02 to 0.08 per cent (Chevalier and Perrot, 1911; Kolanuts and kola chocolate, 1953 Paula, 1938)) and betaine (Mascre and Paris, 1946). Whether these two alkaloids contribute significantly to the physiological action of kolanuts is not known.

Also the mineral composition of kolanuts has been studied. The average contents of various nutritionally important minerals in kolanuts are listed in Table 20.

8.4 KOLA AND ITS ENVIRONMENT

8.4.1 Climatic Requirements

The intensity of kola cultivation increases considerably from the high rainfall (> 1500 mm p.a.) to the lower rainfall areas (1200 to 1500 mm p.a.) within the kola belt. It is in the latter areas that surpluses which enter the kola trade are produced.

As kola penetrates further north into the savanna areas, its cultivation becomes restricted to the particular environment of forest outliers and kurumis (moist stretches along river banks), extensions of the rain forest along streams, swamps and ravines. The environmental conditions in these pockets of forest are quite precarious. The major problem is the duration and intensity of the dry season which may last upwards of 5 months, with very low relative humidity. Kola productivity in this type of environment is low unless there is supplemental water through ground water or irrigation. The supply of ground water depends mainly on the nature and structure of the subsoil, the elevation of the forest outlier in relation to the surroundings and the availability of springs.

The productivity of mature kola trees in West Africa was estimated at 500 kg per hectare (Russell, 1955). Figures from experimental plantings in Ghana have confirmed these yields which are achieved from the fifteenth year onwards.

Kola trees react to moisture changes in their environment by manipulating the leaves. Although the slightly leathery leaves may be retained for more than one calendar year and therefore the tree appears evergreen. When water stress occurs, the tree usually sheds part of its lower leaves thus adjusting its foliage to strike a balance with reducing relative humidity and water availability. During severe droughts all leaves may be shed! Young trees will not survive this type of shock and as a result they require shade for the first two years after transplanting in the field.

The adaptation of kola to the wet rain forest areas is reflected in the mode of seed dispersal and in the characteristics of the seeds.

When the fruit matures, the follicles usually split, but retain the seeds. At this time they are easily separated from the fruit stalk. This may happen incidentally through the action of winds, rain, or purposely by animals or harvesting farmers. When dropping the follicles usually still retain the seeds. The seed coat turns brown and decays in the course of a few days and the kolanuts become exposed to the air. Unless these nuts are protected from drying, they will soon lose their viability, as they are 'wet' seeds. This is especially of importance, because germination occurs very irregularly and part of the seeds may take several months to germinate. This delayed germination is an advantage in carrying the seeds over from November, the time of maturation, to April, when the subsequent rainy season begins.

In its natural habitat the follicles of kola will drop in the leaf litter, where the seeds will be protected against the sun by the canopies of the forest trees and furthermore they will soon be covered by additional litter material. Thus the seeds are preserved until environmental conditions are favourable for germination.

Kola grows well in tropical lowland rain forest areas with temperatures fluctuating around 25 °C, with diurnal ranges of up to 18 °C and frost-free nights, a rainfall of 1250 mm or more, and a dry season of three or fewer months with a dry season precipitation of below 50 mm. The kola tree, as the cacao tree, is adapted to the alternating dry and wet season by its ability to regulate, within limits, its water relationships by partial shedding of its leaves. The nature of the seed (nut), the irregular germination and the requirements for shade during early growth are examples of the tree's adaptation to its environment.

8.4.2 Soils

Kola will respond well to fertile soils with a high organic matter content. However, kola is more tolerant of lighter and less fertile soils than cacao. Kola appears to be more tolerant of less favourable soils also than several other tree crops. These considerations should not lead to siting a new kola plantation in an obviously unsuitable area. It is, however, a matter of discretion whether available fertile areas are allocated to kola or to an alternative crop.

Kola demands a deep, well-drained soil. Badly drained sites are not suited.

Any site selected for kola should have a deep soil profile into which the roots can penetrate and which allows the retention of water in the subsoil either by virtue of its depth or by its silt content. A shallow soil on decomposing rocks, close to the surface, deep soil profiles with a hardpan close to the surface, areas liable to flooding for appreciable periods (expressed in weeks rather than in days) and very clayey soils likely to become waterlogged are all unsuitable for kola.

8.4.3 Overhead Shade

Kola trees are mostly planted in forest areas, where overhead shade for young kola plants can be provided by retained forest trees. However, just north of the forest zone kola may also be cultivated, if the site chosen for its establishment has a cover of savanna trees, of which the canopies are touching, thus providing a light, evenly distributed shade. Without a natural shade, establishment of kola becomes almost impossible in such areas.

8.5 AGRONOMY OF KOLA

8.5.1 The Nursery

8.5.1.1 Raising of seedlings

Kola is generally propagated by seeds. Seeds are often sown at stake. However, sowing of seeds in the nursery usually gives better seedlings. In this case seeds should be pregerminated in order to obtain groups of evenly developing seedlings.

In pregerminating, kola seeds – fresh or stored after the removal of the testa – are sown in seed boxes filled with topsoil mixed in equal parts with fine sand. Seeds are sown in the pregermination medium at a depth of 3 to 5 cm. The seeded boxes are then watered as often as needed; overwatering is to be avoided. Under these conditions stored nuts usually give slightly higher germination than fresh nuts. Germination is usually completed within 80 days in *C. nitida* and 60 days in *C. acuminata*. The pregerminated seeds are planted in baskets or polypots filled with topsoil, at a depth of 7 to 10 cm. They are adequately watered and kept under shade to develop in the nursery. The seedlings will reach transplanting size in six to eight months.
The following factors may affect the development of kola seedlings:

1. *Placement* Placement of seeds in the germination medium affects the formation of roots. Horizontally placed seeds usually result in a higher number of seedlings with straight roots.

2. *Shade* Seeds germinate and seedlings grow better when shaded.

3. *Planting depth* Depth of planting affects germination and seedling development. Plant depths of 5 to 10 cm in the baskets, polypot and seed beds and of 3 to 5 cm in the pregerminating seed boxes are recommended.

4. *Sowing media* Fertile topsoil is the best medium for raising kola seedlings.

5. *Condition of the nuts* Fresh nuts are known to result in slightly better developed seedlings than stored nuts but this has been shown not to be significant enough to discount their slow germination rate.

6. *Size of nut* This does not affect the rate of germination, but it affects the rate of development of seedlings after germination, the bigger nuts usually give bigger and better developed seedlings.

7. *Temperature* Temperature is very important in the pregermination and subsequent development of kola seedlings. Kolanuts germinate best at temperatures ranging from 32 °C to 34 °C. Continuous exposure of seeds to temperatures above 38 °C is known to be lethal. It is also known that although short exposure of seeds to high temperatures in the range of 36 °C to 38 °C speeded up germination, the seedlings which develop from high temperature treated kolanuts tend to be retarded in their growth.

8. *Illumination* Kolanuts stored in transparent plastic bags often turn green and start to germinate, while those stored in black polybags or in the traditional way in wraps of several layers of leaves remain dormant. Experiments have shown that illumination promotes the germination of kolanuts provided adequate humidity is maintained round the nuts.

Finally, the choice of healthy, fully mature, heavy seeds is essential for obtaining well developing seedlings. The speed of germination is variable and can be influenced by several environmental factors.

Pregermination of kolanuts results in a much more rapid emergence than planting of the nuts directly into nursery beds. The seedling should be provided with shade and much water in order to develop within five months to a size which will ensure success in transplanting.

8.5.1.2 Raising of rooted cuttings

Cuttings for vegetative propagation are taken from new growth, which has just hardened, approximately two months after flushing. The time of flushing, therefore, determines the time that cuttings can be collected. Many cuttings will become available in the beginning of the year and smaller numbers throughout the year always with a marked periodicity for any one tree, depending on its pattern of flushing. As ramets established from cuttings should be retained in a nursery for at least 9 to 12 months before they are ready for transplanting, at least 12 months will lapse from taking the cutting up to the subsequent suitable date of transplanting.

8.5.1.2a Choice of material

Kola trees which have given high yields over a long period may warrant vegetative multiplication. Other factors to be taken into consideration are: size of the tree, average nut size, distribution of the production over the year and the occurrence of biennial fruiting cycles. However, productivity will usually be the first determinant.

Cuttings can only be collected from trees which have healthy, well-developed flushes. At two to four months, the leaves of the new flushes have become dark green and hardened. By this time the stem has taken on a dull dark green or brownish colour. With cuttings taken at this stage the best results are obtained in propagation. Later, when the stem has lost its epidermis and much secondary growth has occurred, the success in rooting becomes less. Cuttings with thick, dark brown stems and senescent leaves, often overgrown with lichens, cannot be rooted successfully. This is confirmed by Clay (1963a) and Dublin (1965).

8.5.1.2b The propagator

Attempts to root kola have either been made by placing cuttings in concrete propagators or in wooden boxes covered with polythene.

Swarbrick (1964) summarized experimental work carried out on the rooting of kola cuttings at Ibadan. He showed that success in rooting partly depends on a strict maintenance of a maximum humidity of the air surrounding the cuttings and on a low light intensity. Both factors cooperate to reduce transpiration from the cuttings.

It may be possible to achieve these conditions also in less permanent structures: wooden propagators or, on a small scale, wooden boxes with a height of 60 to 75 cm, allowing sufficient space for the rooting medium and for the cuttings to be planted.

The propagator is first laid with a layer of 2.5 cm of washed gravel or granite chips. A thick layer of moist sawdust, about 30 cm deep, is placed on top of this, then levelled and firmed. The sawdust should be devoid of foreign particles and changed once every six months at which time the propagator should also be properly cleaned.

8.5.1.2c Equipment and materials

The following items are required at the nursery when propagating kola through cuttings:

1. *Secateurs* A secateurs is required for taking the cuttings.

2. *Knife* A sharp knife is required to cut and give the base of the cuttings a clean surface. It may also be used to take the cuttings in the absence of a secateurs.

3. *Bucket* a receptacle, such as a bucket, is necessary for carrying the cuttings.

4. *Polythene bag* Polythene bags are convenient containers to supplement buckets for keeping the cuttings in a fresh state. The cuttings should be put in the bag, which is closed above them.

5. *Headpan* A headpan in which cutting are placed and finally cut before setting is also required.

6. *Baskets* Baskets, 30 cm high and 18 cm wide, woven with the midribs of oil palm fronds are needed. Larger baskets are later required when the smaller baskets deteriorate. Polythene pots are not suitable for potting, because they are not rigid enough, and may allow the fragile roots to be damaged by sagging, but if baskets are not available, polypots can be used provided they are strengthened with sticks.

7. *Potting medium* Topsoil collected from forest should be used for potting, as such soils are rich in humus.

8.5.1.2d Rooting procedures

Cuttings are collected when it is cool and humid, preferably in the mornings or later in the day if the weather if overcast. Moisture loss from cuttings should be prevented. If the material is collected from trees close to the propagator, it can be enclosed in polythene bags. If the cuttings are to be transported over long distances, they should be placed in a bucket, with their bases in water, and covered with polythene. It should be attempted to bring the cuttings rapidly to the propagator.

The cuttings should be 15 to 20 cm long and the base should not include a part of that flush from which the selected flush emanated. The few lower leaves are removed, together with any flower buds. The apical tuft of six leaves or more is left. Subsequently, the cuttings are cut at the base with a sharp knife, under water, to provide a fresh clean surface and then they are placed, obliquely, in the rooting medium to a depth of 10 cm.

Immediately after setting the cuttings are watered intensively to allow the medium to settle firmly around their bases, and the propagator is closed carefully. On subsequent days the cuttings are sprayed thrice daily with a knapsack sprayer until a film of water appears on the leaves. If this is not done conscientiously, the cuttings will quickly drop their leaves and die back. The other extreme, overwatering, causes waterlogging, poor aeration and rotting of the bases of the cuttings. This has to be prevented too.

Apparently Dublin (1965) halved the leaves of his cuttings. He also used cuttings only 10 cm long, retaining two halved leaves. It is not clear whether he

Table 21 Success percentages in rooting of cuttings taken from various clones, at three months from setting

Success percentage in rooting	Number of clones	Number of cuttings per clone
1–10	10	186
11–20	14	295
21–30	12	206
31–40	10	205
41–50	4	70
51–60	6	361
61–70	4	140
71–80	2	50
81–90	1	28

used only apical cuttings as described above, or also lower placed stem cuttings.

Mature trees differ widely in their capacity to be propagated by means of stem cuttings. Percentages of cuttings from 63 different trees successfully rooted at three months from setting at Ibadan during 1964 and 1965 are presented in Table 21 (van Eijnatten, 1966, 1969).

The average percentage of success in rooting (28 per cent was somewhat lower than that reported by Dublin (1965), which was 34 per cent. The difference may be explained by the time limit of three months after setting of the cuttings, as used at Ibadan, Nigeria.

Bodard (1962) and Clay (1963) found that cuttings from juvenile kola material (seedlings, suckers) gave higher success percentages than those from mature trees. But, because such trees have not yet shown their yielding ability, this does not help in establishing promising clones.

8.5.1.2e Potting of rooted cuttings

The roots start to develop from callus tissue at the base of the cutting, on average nine weeks after setting. There is usually only one thick, stout, but fragile root, sometimes two and occasionally more, up to five or six. These roots do not ramify, initially. The roots often arise at an angle of up to 90 °C. Cuttings with roots deviating more than 45 ° should be discarded, because plants developing from such cuttings are not firmly anchored in the soil, especially if only one main root was formed. In the second or third year, the ramet will have formed many new adventitious roots from its stem, radiating in all directions, after which time the plant becomes less dependent on the original root system for anchorage.

Successfully rooted cuttings are potted in rigid containers at least 40 cm high in order to provide sufficient space for the new root. A rigid container is mandatory. In any flexible pot the newly formed roots are likely to break, as they are very brittle. At Ibadan, Nigeria, the cuttings are successfully potted in baskets, just like kola seedlings, measuring 50 cm high and 25 cm in diameter. Topsoil from forest areas is used as a potting medium.

Some cuttings show yellowing leaves at the time of potting, especially when rooting has taken a long time. If this is the case, it may be advisable to use a mixture of compost and topsoil or topsoil enriched with a low dose of nitrogen-containing fertilizers.

If potting is unsuccessful due to damage to the roots or otherwise, this will become apparent between the first and the third week after potting. However, this does not often occur, contrary to experience in the Central African Republic, where Dublin (1965) observed a high percentage of die-back after potting. He blamed this mainly on the necessity of cutting the newly formed roots at the moment of transplanting. With the use of sufficiently high, rigid containers this should be unnecessary. Dublin, however, rightly suggested that a method should be found to root kola cuttings directly in pots, so that potting of newly rooted cuttings may be avoided. The utilization of pots with a central core of sawdust surrounded by topsoil or compost is suggested.

Immediately after potting, the newly rooted cuttings are transferred to a propagator and watered thoroughly. The propagator is then closed and for the following three weeks only opened for a rigid inspection of the plants. In this period the ramets can settle in their new baskets. Periods of hardening shorter than three weeks invariably cause a high percentage of die-back. From the third week onwards, the polythene covered frames are gradually lifted in the course of one week. The plants have then become sufficiently settled and accustomed to a normal relative humidity, so that they can be transferred to a shaded nursery. They are conveniently placed on beds 1.20 m broad and separated by paths of 50 cm.

When the ramets have been placed on the nursery beds, they should be kept watered just like the seedlings. A light shade, intercepting approximately one-third of the light, is desirable. Exposure to full sunlight will soon cause drop of leaves and stunted growth. Routine maintenance procedures should be applied as in the case of potted seedlings.

Although no deficiency symptoms have yet been observed in rooted cuttings kept in nurseries at Ibadan, it is conceivable that they may occur, especially when the ramets have to be kept within the confined soil space of the baskets for a period of up to two years. Application of fertilizers may promote the growth of such plants.

8.5.2 Establishment of Kola

8.5.2.1 Land preparation

In the past, much of the kola was cultivated more or less incidentally, as it was protected and nurtured whenever it happened to germinate. Preparation of the land only became of importance when kola was planted in continuous plantings. Russell (1955) reported that the Western Nigerian farmer completely clears the ground prior to planting kola. In fact, kola is often planted in newly established food farms. During the few subsequent years of cultivating cassava, maize, yams or cowpeas, the weeds are kept in check. So the young kola tree is given a chance to develop prior to reversion of the land to bush, after which only an occasional slashing of the weeds around the trees is practised. However, an exposure to full sunlight leads to a slow initial development of the tree. Several other methods of establishing kola farms are, therefore, being tested at Ibadan. Only the results of the first few years are available, but they are informative (van Eijnatten, 1966a).

There are three methods of land clearing for planting kola. These are clear-felling, selective thinning and simple felling.

The growth of seedlings and ramets is influenced appreciably by the various methods of establishment. The ramets grow taller in the simply felled area. Seedlings tend to follow suit in this respect. However, differences between the various establishment methods are much more reflected in the general habit of the plants and in the occurrence of diseases.

The fully exposed seedlings in the clear-felled area have a compact habit, short internodes, and are frequently branching. They suffer from various leaf spots and die-back of the twigs. The sheltered seedlings in the selectively thinned forest have an etiolated appearance and large leaves, often covered with lichens. The seedlings in the simply felled area are less compact than the exposed seedlings, harbour less leaf spots and generally look more lush and vigorous. The same is true for the ramets, which develop best in the half-sheltered environment of simply felled land. The provisional conclusion is that kola trees established on clear-felled land show an undesirably compact growth and may be more liable to attack by various fungal diseases than sheltered trees.

The less expensive method of establishing kola in thinned forest apparently admits insufficient light to the young trees. If the thinning is done more thoroughly, the method will come close to that of simple felling both in cost and in performance of the trees. Simple felling and a temporary regrowth of the vegetation between the traces along which the kola is planted, appear to result in the best developing plants.

8.5.2.2 Spacing

Desirable spacings can currently only be estimated from the appearance of kola plantations established at regular spacings and irregularly planted farmers' kola. A large-scale experiment to test a range of spacings from 1.50 to 10.50 m was initiated at Ibadan during 1965, but wider spacings for mature trees may have to be combined with a system of planting at close spacing to ensure higher yields per hectare during the early productive years of the crop. This should be followed up by timely reduction of the number of trees to allow for a good development of the ultimate stands.

It is suggested that mature trees should be spaced at 7.50 to 9 m depending on the fertility of the site. This is true for seedlings and ramets. Interplanting of extra seedlings can be done halfway between the ultimate stands, both within the rows of trees and on the diagonals. This reduces the spacing to 3.75 or 4.50 m. It serves no purpose to establish seedlings at closer spacings, as in that case thinning may become necessary prior to the moment at which the trees reach their productive age. Reduction of the number of trees may have to start around the tenth year.

Ramets, which are productive at an early age, can be interplanted at closer spacings, e.g. 2.50 to 3.00 m. The reduction of the number of plants has to be started earlier than with the seedlings, probably in the sixth year, because of these shorter planting distances. At this time ramets may already have been in production for four years.

Results will not be available for several years. A free growing tree forms a wide, dome-shaped canopy, with foliage often reaching to ground level. Dublin (1965) suggested that the diameters of the canopies of such trees reach 6 m in the tenth year and 9 m in the twentieth year. For Nigeria the average canopy diameter of mature trees is around 8 m. The ideal habit is often achieved by trees growing in compounds or other isolated spots. Many of the kola farms, however, have been planted at narrow spacings varying from 4.50 to 6 m. When the trees grow up with these narrow spacings, the canopies soon touch and the trees force each other to become tall. Such trees are etiolated and have a thin foliage at the ends of the branches. The lower limbs are lost and the trees become less accessible for harvesting. The effect of the narrow spacing is similar to that of the retention of a too dense overhead shade.

Only Dublin (1965) has so far provided figures on differences in productivity due to variations in spacing. He found that trees planted at 5 × 5 m yielded 125 nuts per tree when nine years old, giving 49,815 nuts per hectare. Trees planted at 8 × 8 m gave 150 nuts per tree or 20,685 per hectare. So wider spacings allowed higher yields per tree, but also reduced the total yield per hectare.

8.5.2.3 Planting operations

8.5.2.3a Time of planting

Newly planted material, whether seedlings or ramets, has undergone a thorough change of its environment and is therefore vulnerable. The one factor which cannot possibly be controlled when planting on a large scale is the availability of water. So the young plants are brought to the field when sufficient rains have fallen to ensure the availability of soil moisture and when, furthermore, the rains have become reliable. A prolonged dry spell occurring just after planting may be disastrous for the young plants. A suitable planting time may be in April for the western part of the West African kola belt, in most areas, however, it is May. North of the kola belt, it may be necessary to postpone planting till June.

The later in the rainy season the plants are established in the field, the less prepared they will be to withstand the ensuing dry season. It is advisable to allow a full half year of growth under moist conditions, whenever possible. Plants established very late in the rainy season often die during the subsequent dry season, unless provided with ample water.

The use of Kolanuts, instead of seedlings or ramets, as planting material allows even less time for establishment of the young plants, especially if slowly germinating Kolanuts were used. Some time may be gained if the largest possible nuts are selected for planting and split prior to sowing. This would ensure an early germination and a reasonable supply of reserves from the (large) cotyledon. The only advantage of a rapidly germinating Kolanuts sown at stake, above transplanted material, is that the taproot will not have to be disturbed as in transplanted material, which had its roots cut at a depth of 30 to 40 cm.

8.5.2.3b Shade requirements

Shade requirements of seedlings and ramets were studied at Ibadan. Nigeria (van Eijnatten, 1966b). Various shade densities (from 0 to 66 per cent shade) were tested. The required level of shading was obtained by the construction of slat houses. The 5 cm wide slates were placed at a distance of 10 cm, 5 cm or 2.5 cm from each other, thus intercepting respectively 33, 50 or 66 per cent of the light. Plants were kept free from water stress by regular watering throughout a year. Seedlings and ramets (cultivar: Agege Red) were placed in the slat houses when they were six months or one year old, the normal ages at which the two types of planting material are transferred to the field.

In a first experiment Agege Red seedlings showed a very beneficial effect of the presence of shade on the height of the seedlings, their dry matter content, the number of the leaves and their average length (Table 22). The shaded

Table 22 Observations on kola seedlings after one year under various shade regimes. The experiment started in November 1963 and was terminated in December 1964

	Shade percentage		
Observations	0	33	66
Height (cm)			
November 1963	14	14	11
December 1964	25	40	71
Dry weights (g. per. plant) (December 1964)			
Roots	7.6	26.0	16.7
Aerial parts	13.4	63.6	66.8
Stems	9.3	20.1	21.9
Leaves	4.1	43.5	44.9
Total	21.0	89.6	83.5
Leaves (December 1964)			
Number	13	39	44
Length (cm)	11.5	12.6	14.8

Table 23 Estimated leaf areas (cm^2) of Agege Red kola plants raised under different shade regimes

Seedlings			
Time from transplanting (months)	Shade percentage		
	0	33	66
1	223	178	116
4	189	320	350
6	317	775	518
12	406	1470	2311

seedlings increased in dry matter to approximately 4 times that of unshaded plants. The plants under a light shade (33 per cent) produced slightly more dry matter than the plants kept under 66 per cent shade. There was a decreasing relative development of the root system with increasing density of the shade. Similar findings were made in a second experiment in which ramets from three kola clones were raised under 50 per cent shade or without any shade.

Ramets responded to shade in the same way as the seedlings. The plants were taller and rooted more deeply under shade than without shade. The shaded ramets produced almost double the quantity of dry matter and number of leaves as exposed ramets.

Estimates were made of leaf area on seedlings and ramets in the two shading experiments (Table 23). Soon after transplanting, the seedlings placed in full sunlight shed some of their leaves in spite of a regular supply of water. Shaded plants did not show a reduction in leaf area after transplanting. Initially, seedlings under light and dense shade regimes kept pact with each other in development of their leaf area. Subsequently, however, the densely shaded plants increased their leaf area far more rapidly than the plants which got 33 per cent shade. After 12 months the densely shaded plants had 50 per cent more leaf area than the lightly shaded plants. The exposed seedlings had a much smaller leaf area than the shaded seedlings.

The practical implications of these data are, firstly, that young plants, whether seedlings or ramets, need shade during their early development and, secondly, that they can be overshaded. A dense shade reduces the relative development of the roots in seedlings. This is a great disadvantage once the tree outgrows the shade or when shade can no longer be provided economically.

However, already in its second or third year the kola tree requires light for the formation of a broadly branching canopy with a well-developed foliage. Provided the requirements for moisture are fulfilled, exposure to full sunlight is favourable for the growth of the mature trees.

8.5.2.3c Transplanting seedlings

Transplanting seedlings from the nursery to their permanent site will be most successful if it is carried out early in the rainy season. The plants then have the longest possible period to become settled prior to the subsequent dry season.

At transplanting, the seedlings should preferably be 40 to 50 cm high and carry 12 to 14 expanded leaves. At this stage, the root contains one-fifth to one-quarter of the total dry matter of the seedlings (van Eijnatten, 1966). If the seedlings are smaller than 20 cm and carry less than six leaves, they are not suitable for transplanting, because many of them die back gradually or remain stunted after transplanting. So it is important to select only well-developed specimens with a good number of leaves for transplanting purposes.

Seedlings raised in seedbeds are generally transplanted with bare roots. The following procedure is recommended: a furrow should be dug across the nursery bed to a depth of 30 to 40 cm. From the side of the furrow, the plants should be worked loose with a trowel or garden fork, taking care not to damage the roots. The taproot should be cut at the bottom of the furrow. The plants are sprinkled with water and then covered with wet sacks or placed in polythene bags. The process is best done on a rainy day.

Four- to six-month-old seedlings still retain turgid cotyledons. It is essential to take care that these are preserved on the plants, because they serve as a food reserve for the seedling. Careless or rough handling during uprooting and transplanting may cause the cotyledons to be severed from the plants. The presence of the cotyledons on seedlings of six months apparently helps the plants to endure the shock of transplanting, especially when the seedlings are uprooted from the nursery beds and transferred to the field with bare roots. Later, when the seedlings become nine months or one year old, the cotyledons shrivel and drop.

The time from uprooting to planting in the field should be kept as short as possible by preparing the planting holes in the field prior to uprooting the seedlings from the nursery. The holes should measure 30 by 30 cm with a depth of 30 to 40 cm. the subsoil removed from the hole is to be deposited in a heap at a distance from the hole. This will make it possible to scrape the topsoil around the planting hole into it after the seedling has been planted. The soil should be thoroughly firmed around the root of the kola seedling to achieve an early close contact between the root and the soil, thus promoting the absorption of water. The plant should be placed in the planting hole with the main root pointing straight down and the cotyledons at the level of the surrounding soil surface. If water is easily available, 5 to 10 litres should be applied to carry the seedling over to the next rains. Watering is not necessary if there is sufficient water in the soil or if rain is expected on the day of transplanting.

At the time of transplanting, shade should be provided intercepting approximately one-third of the light. This will prevent wilting and undue leaf drop (van Eijnatten, 1966). The shade can conveniently be provided with three of four oil palm fronds, tied together above the plant (Quarcoo, 1975).

If the seedlings were raised in containers, these should be carried to the field after a final application of water in the nursery. When removing the potted seedling, the taproot should be cut below the pot, if it has grown through it. If the containers were made of easily perishable materials, they can be placed directly into the planting holes, the spaces being filled up with topsoil scraped together from around the holes. The plants should immediately be provided with a light shade, like the seedlings transplanted with bare roots. Sometimes, it may be necessary to transplant seedlings into the field after they have been in the nursery for 9 to 12 months. At this time they have reached a height of 50 to 80 cm, lost the cotyledons and they carry 15 to 20 leaves. Transplanting these seedlings unpruned with bare roots often results in failure. Pruning of the seedlings to a height of 30 cm ensures successful transplanting, but it retards the development of the plant considerably (van Eijnatten, 1966). It is advisable to attempt transplanting such seedlings with a 'ball of earth' (Dublin, 1965), since they have started to produce secondary roots. With care, it is possible to preserve earth around this secondary root system. Seedlings transplanted in this way will be able to continue absorbing water. A better solution would be to pot the seedlings in baskets of 40 to 50 cm deep, having a diameter of 20 to 25 cm, when they are four to six months old. During the remaining months in the nursery the seedlings can settle in these baskets and transplanting can be carried out virtually without stagnation of growth. Also seedlings of 9 to 12 months should be provided with a light shade when they are transplanted.

8.5.2.3d Transplanting rooted cuttings

Transplanting into the field of healthy, vigorously growing ramets, less than 9 months from potting, usually leads to an early death or to a poor, stunted development. This is probably due to the fact that the root system is not yet sufficiently developed at that time. Around the ninth month a noteworthy increase occurs in the relative importance of the root system.

Transplanting ramets from the nursery to the field at less than 9 months from potting usually leads to early death, but the most rapid establishment is obtained with ramets of 18 months or more. This is probably related to the acceleration of growth observed in ramets reaching the age of 18 months in the nursery. However, the older the ramets, the larger the plants, the less easy is transportation. If the ramets are to be carried in motorized transport over appreciable distances, they should be protected with polythene hoods against the rapidly desiccating effect of air draughts, even with humid or rainy weather.

Also in the case of rooted cuttings it is essential to provide the transplanted materials with temporary shade. This is likely to be required during the first two years of development, especially during the dry seasons.

8.5.2.4 Interplanting

The establishment of kola in food farms allows for the land to be used for annual crops during the first few years after establishing the kola. A disadvantage is that for several years the kola trees will be growing on a completely exposed field, virtually without any vegetation to protect the soils and the young kola trees during most of the dry seasons and part of the wet seasons.

As regards kola, this is not a desirable practice, the less so because the effect may still be felt long after intercropping has been stopped. Interplanting various tree crops such as cacao and coffee leads to similar problems in causing the kola trees to be exposed for the first two years after transplanting. The soil may be less exposed than when annual crops are interplanted, because a natural weed cover can be allowed or a legume cover crop established.

Russell (1955) states that cacao and kola are too competitive to be grown together with much success. The productivity of 65 mature kola trees in a 30-year-old cacao farm at Ibadan proved to be low: 155 kg per hectare (van Eijnatten, 1966a). Robusta coffee interplanted between kola may grow and yield well in Western Nigeria (Russell, 1955). This experience has also been gained in the Central African Republic (Dublin, 1965) with kola planted at a spacing of 9 × 9 m and coffee interplanted at a spacing of 3 m. The coffee bushes should be removed gradually to allow a free development of the kola trees. The coffee thus provides the farmer with a return on invested capital long before the kola has started to produce. If the aim is a permanent cultivation of both crops in a combined plantation, the kola trees should be planted more widely, e.g. 12 × 12 m, allowing ample space for single or double rows of coffee between those of the kola trees. The effect of the neighbouring coffee bushes on the development of the kola trees and their (early) productivity is not known.

8.5.3 Post-planting Maintenance

As food crops are often grown between the kola trees during the early years after their establishment, some cultivation is carried out automatically. When annual crops are no longer raised within the kola farms, maintenance becomes restricted to an occasional slashing of the weeds, usually twice a year. Under mature trees weed growth is thoroughly cutlassed to enable the farmers to easily spot dropped fruits. However, kola may respond to a more careful treatment than the one meted out by many a farmer, particularly in respect of shade and weed control, application of mulch, application of fertilizers and sanitation of the trees.

8.5.3.1 Shade and weed control

During the first, and sometimes also during the second year after establishment in the field the kola plants are to be kept under shade intercepting approximately one third of the light. In West Africa this is usually achieved by placing three or four pieces of palm fronds around the plant and tying these together beside and above the plant, unless a natural shade of the desired density is present. The palm fronds may have to be replaced three or four times at intervals of six months, depending on the development of the plant. When the plant has settled well and is in active growth after this period, shade should no longer be provided. Any vegetation within 100 cm from the projection of the kola tree's canopy on the ground should be removed and no overhead shade should be tolerated.

If kola has been established under thinned forest, it is advisable to ringbark some of the shading trees at the end of the second year before the dry season. These trees will then die slowly and during the ensuing rainy season the shade wil have been reduced. It is important to realize that kola must be grown without shade, if it is to attain its full development and productivity.

Especially in newly established kola plantations within which the vegetation is allowed to regrow between the kola plants, regular weeding has to be carried out. Within a distance of 80 to 100 cm from the stem of the young plant, weeds are to be removed by hand in order to avoid damage to the young adventitious roots radiating from the stem, just below the soil surface. Weeds at a greater distance may be removed by means of a hoe.

If, inadvertently, weeds have overgrown the kola plants, or when densely shaded plants are to be given more light through shade reduction, any corrective measures should be taken gradually. For instance, the stems of the weeds covering the kola plants may be cut, but the plants should not be removed. The remnants of the weeds will drop gradually, they will decay or be blown away by the wind. Dense shade can be thinned gradually by ringbarking or pruning the shading trees. If these various measures are carried out suddenly, the kola plants are likely to suffer from severe sunscald, many leaves may drop and the development of the trees may be seriously retarded.

8.5.3.2 Mulch

To be able to protect its own root system with the shade of its canopy, a kola tree must have attained an appreciable size, especially in respect of the radical development of the canopy. It usually also has formed deeply reaching roots by that time and so it depends less on moisture retained in the upper layers of the soil. Up till then it is advisable to provide a large quantity of mulch. The mulch should be applied in a ring covering the area between 15 and 100 cm from the trunk. The gap between the plant and the mulch will allow for a free circulation of air and it may prevent the occurrence of fungal diseases and insect pests on the lower part of the trunk. To be effective, the mulch should be applied some weeks before the end of the rainy season.

Material to be used for mulching is collected from the surrounding vegetation or from nearby grassland. It is important to use only herbaceous material, as decaying wood, especially the coarser parts, may form a substrate for root diseases such as *Fomes noxius*. An effective mulching material is formed by the stems and leaves of banana or plantian. Whatever material is used, it is piled up within the area to be covered to a compact layer of 30 to 40 cm thick. This material should be allowed to decay on the spot. It will enrich the soil and increase its organic matter content. The practice of removing the mulch after the dry season is wrong and can only damage the rootlets of the kola feeding on the decaying mulching material.

8.5.3.3 Plant nutrients

Raising kola seedlings in sand cultures with incomplete nutrient solutions made it possible for Egbe (1966) to obtain well-defined deficiency symptoms for various nutrients. Lack of nitrogen produced its effect within a month from the start of the experiment, while the effects of lack of phosphorus, calcium and magnesium occurred later. The seedlings showed no reaction to a deficiency of potash after 18 months. Egbe's descriptions of deficiencies are given below:

1. *Nitrogen deficiency* The seedlings become stunted and all plant parts become pale yellowish green, starting with the older leaves and spreading to the young ones. Growth is retarded and fewer leaves are produced. The root system develops weakly.

2. *Phosphorus deficiency* The edges of young leaves fold downwards. The leaves become less flexible. The lower leaf surfaces become bronze-coloured. The bronzing spreads to the petioles and stem tips. Die-back may occur from the leaf tips and the tops of the plants.

3. *Calcium deficiency* At the leaf tip and the margins in the top third or half of the leaves a light-brown scorch occurs. The lower portions of the leaves do not even show a chlorosis. Leaf size and length of internodes are reduced, the plant becomes stunted.

4. *Magnesium deficiency* Brown, necrotic spots occur between the veins, starting from the edges of the oldest leaves. The spots spread inwards, but the leaf tissues surrounding the midrib and the veins remain green. Affected leaves drop and the seedling retains ultimately only the youngest leaves.

These descriptions may be helpful in discovering gross deficiencies of various nutrients in seedlings or ramets, especially those which are raised for an appreciable time in the same confined root space of a basket.

Experimental work on the application of fertilizers to plants after their transfer to the field has been initiated. During the first year two applications of 20 g of sulphate of ammonia and 10 g of both superphosphate and potassium sulphate per plant followed by three applications, during the second year, of 50, 25 and 25 g respectively did not influence the growth of the seedlings or ramets in Nigeria. Applications 1.5 to 2 times as high still had no influence at Kade, Ghana. Again in the case of young plants, transferred to the field, the only recommendation as regards the use of fertilizers is that these should be applied only whenever deficiency symptoms are observed, until such times as more information regarding the influence of fertilizer dressings on kola trees will be available.

8.5.3.4 Sanitation of the trees

The effect of pruning kola trees has not yet been studied. Pruning may help in reducing the ultimate size of the trees, which would facilitate pollination, harvesting and control of pests and diseases. Bodard (1955) considered pruning harmful, because he found that removal of parts of branches often leads to a regressing die-back.

It is advisable to keep the trees in a state of sanitation by removing dead wood which is often found within the canopies. This should be done thoroughly by cutting back into living tissues and painting the cut surfaces to prevent infection of the wounds by fungi. Trees seriously damaged by stem borers (*Phosphorus* spp., **Coleoptera**) are often better regenerated from the coppiced trunks, which are free from borer damage. Less seriously damaged trees can be partially regenerated.

Kola trees are often parasitized by various plant species, which should be removed regularly. Common parasites are various loranthaceous plant species. Their removal involves cutting of branches. The wounds thus caused should be painted. It is also desirable to remove the frequent saprophytes such as orchids and ferns, flourishing in crevices of the bark, at the junctions of trunk and branches, in old wounds and so on.

8.5.3.5 Regeneration and rejuvenation

Unlike several other tree crops, such as cacao and coffee, kola does not regularly produce chupons or suckers. Only occasionally a tree may be found to produce some chupons, but these arise from horizontal limbs of the tree, often at a considerable height. So they are not useful for the regeneration of the tree. However, trees may have become too old, too tall and overgrown or they may have been seriously damaged by stem borers. In these cases it is desirable to replace the old trees with young material, developed from the old trunks. Regeneration or rejuvenation could also be useful in many too closely planted kola groves. By selecting from the old stands trees at a desirable spacing and regenerating these while removing the remaining trees, a virtually new orchard may be established in a few years.

Usually many shoots develop from suborbicular outgrowths on stumps of coppiced trees; they form a sort of cushion or swelling, carrying many buds at its apex. These outgrowths can attain a diameter of 4 to 5 cm and apparently they arise from phellogen tissues in the bark. The abundance of young growth on recently coppiced trees attracts insects. These shoots can rarely develop undisturbed, unless protected by a regularly applied insecticide (DDT or BHC). The shoots should be thinned out to two or three, implanted with a healthy and strong junction low on the trunk and spaced more or less evenly around it.

An experiment in Nigeria showed that 93 per cent of the trees coppiced at 120 cm from the ground gave rise to cushions bearing shoots; for those coppiced at 30 cm and 60 cm the percentages were 73 and 78 per cent respectively. The remaining trees died back within some months from coppicing. The time of coppicing the trees (at the beginning, the middle or the end of the rainy season) had no influence. The number of cushions formed per tree and their diameter (measured from the trunk to the furthest point) was influenced by the height of coppicing (Table 24).

Table 24 Influcnce of height of coppicing kola trees on the formation and growth of callus tissues from which new shoots may arise

Height of coppicing	Percentage of plants producing cushions	Number of cushions per tree ($P<0.1$)	Diameter of cushions (cm) ($P<0.05$)
30 cm	73%	35(± 13)	3.7(± 0.1)
60 cm	78%	57(± 13)	3.7(± 0.1)
120 cm	93%	98(± 13)	3.2(± 0.1)
Average	—	63(± 7)	3.5(± 0.1)

In all respects it is desirable to coppice the trees at 120 cm. The height is no disadvantage, because the newly developing shoots start branching very rapidly. Their high position may make them more prone to insect attack, but when regularly dusted with an insecticide, the shoots should have the advantage of not having to complete with weeds.

Although the time of coppicing does not influence the number of cushions formed per tree, it seems advisable to coppice trees at the beginning of the rainy season. This will allow the newly formed shoots to go through their first months of development in a relatively cool and cloudy weather and, when the dry season seats in, the young shoots and leaves will have become hardened.

The growth of coppiced trees is rapid and production may be resumed in the second or third year from coppicing. At Agege, Nigeria, shoots on coppiced trees took only a year to attain a height of approximately 100 (± 16) cm, measured from the junction with the trunk. In another regeneration experiment at Kande, Northern Nigeria, regrown trees reached a height of up to 6 m after five years.

8.6 DISEASES OF KOLA

Many diseases of kola have been recorded. Only in a few instances, however, is information available on economic importance and on methods of control.

8.6.1 Fruit and Nut Diseases

Kolanuts are stored in baskets lined with fresh leaves; this is often done very soon after the harvest. At this time the transpiration rate is still very high and droplets of water may form on their surface. This provokes the development of various parasitic fungi: *Diplodia macropyrena* Tassi, *Fusarium solani* (Mart.) Sacc., *Fusarium moniliforme* var. *subglutinans* Wollenw. et Rg., *Gliocladium roseum* (Link.) Bain., *Penicillium* spp., *Pleurotus colae* Massee and *Schizophyllum commune* Fr. Especially the wet rots caused by *Fusarium* and *Penicillium* species are common. Infection of nuts can be prevented, when the nuts are allowed to attain their restive stage, during which transpiration is greatly reduced, prior to storage.

Another disease of kolanuts is caused by *Botryodiplodia theobromae* Pat. This fungus infects the follicle, which develops a black rot, and subsequently attacks the nuts. Rusty brown spots appear on the nuts, which later turn black and become hard and dry. The dead tissues may fall out, leaving small pits in the surface.

A *Phytophhora* sp. was recorded on young kola fruits in Ivory Coast (Bodard, 1955).

Figure 29. Photograph of leaf diseases of kola

8.6.2 Leaf Diseases

A leaf disease common on exposed kola trees at Ibadan, Nigeria, is shown in Figure 29. Only immature leaves are attacked, usually in the later part of the rainy season. The first symptoms are brown, angular spots, especially at the tips of the leaves. Later, the leaves start to die back from the tips and finally they drop. Seriously affected kola plants are left with many leafless shoots. The loss of leaves induces the plant to flush vigorously, but again the new leaves may be affected and drop. This results in a bushy appearance of the plant. The responsible fungus has not yet been identified conclusively. *Pestalotia* and *Glomeralla* species have been suggested.

8.6.3 Root Diseases

Two root diseases, *Fomes lignosus* Klotzsch and *F. noxius* Corner, have been recorded on kola. At Ibadan these diseases occurred only on kola planted in clear-felled land. Attacked plants suddenly drop their leaves and die one or two weeks later. The roots of such plants are covered with brown rhizomorphs and fruiting bodies may be formed at the base of the trunk. *Fomes* attack can only be prevented by thorough removal of logs, stumps and roots from the area where kola trees are to be planted.

Mallamaire (1954) reported two more root diseases of kola respectively caused by *Leptoporus lignosus* (Klotzsch) Heim ex Pat. and *Sphaerostilbe repens* B. et Br.

8.6.4 Other Diseases

Swollen shoot virus of cacao has been identified in *Cola chlamydantha* and *C. cordifolia* (Todd, 1951). Apparently *C. nitida* is also a carrier of this virus. At Ibadan several *C. nitida* seedlings have been found with marked swellings of the stem, reminiscent of the classical symptoms caused by swollen shoot virus in cacao. Preliminary attempts to transmit the virus from these plants to kola or cacao seedlings were unsuccessful.

8.6.5 Summary of Fungi Recorded on Kola

Fungi recorded on kola are listed in Table 25 with details where available.

8.7 PESTS OF KOLA

8.7.1 Kola Weevils

Kola weevils are a major pest of kolanuts, both prior to the harvest and during storage. Several curculionid weevils attack the nuts *Balanogastric kolae* Desbr.

Table 25 List of diseases associated with kola in West Africa

Fungus	Country	Host	Common name	Parts of plant affected
Botryodiplodia theobromae Pat	Ghana	*C. nitida* *C. acuminata*	Small pox	Nuts
Diplodia macropyrena Tassi	Sierra Leone	*C. nitida*	—	Nuts
Fomes lignosus Klotzsch	Ghana	*C. nitida* *C. acuminata*	Root rot	Roots
Fomes noxius Corner	Ghana	*C. nitida* *C. acuminata*	Root rot	Roots
Fusarium solani (Mart.) Sacc.	Ivory Coast	*C. nitida*	—	Nuts, especially on embryo (yellow mycelium)
Gliocladium roseum (Link.) Bain.	Ivory Coast	*C. nitidia*	—	Grey mycelium around embryo
Leptoporus lignosus (Klot.) Heim ex Pat.	—	*C. nitida*	—	Roots
Marasmius byssicola Petch	—		Thread blight	—
Marasmius scandens Massee.	Ghana	*C. nitida* *C. acuminata*	Thread blight	Canopy
Marasmius trichorhizus Speg.	—	—	Thread blight	—
Melanconiopsis africana	Ivory Coast	*C. nitida*	—	Branches
Melomastia hyalostoma Luc.	Ivory Coast	*C. nitida*	—	Branches
Micopeltis depressa Cooke et Massee	Fernando Po	*C. acuminata*	—	Leaves
Nectria fulliginosispora Luc	Ivory Coast	*C. nitida*	—	Branches
Pellicularia salminicolor (B. et Br.) Dort (= *Corticium javanicum* Zimm)	Indonesia	*C. nitida*	—	Branches, trunks
Penicillium spp.	Ivory Coast	*C. nitida*	—	Nuts
Phaeobotryosphaeria varians Luc	Ivory Coast	*C. nitida*	—	Branches
Phytophthora spp.	Ivory Coast	*C. nitida*	—	Young fruits
Pleurotus colae Massee	Ghana	*C. nitida*	—	Nuts
Sphaerostilbe repens B. et Br.	—	*C. nitida*	—	Roots
Schizophyllum commune Fr.	Ivory Coast	*C. nitida*	—	Nuts
Teichosporella pachyasca Luc.	Ivory Coast	*C. nitida*	—	Branches

and at least five *Sophrorhinus* spp. (*S. divareti* Hoffm., *S. imperata* Fst., *S. pujoli* Hoffm., *S. quadricristatus* and *S. simiarum*) (Alibert and Mallamaire, 1955; Caswell, 1962; Pujol, 1962).

These weevils are reported to attack wounded or damaged fruits. Eggs are laid in the nuts or in other parts of the fruit. The period from laying to the emergence of the adult weevil is approximately one month. *Balanogastric kolae* has an average adult lifespan of 53 days, *Sophrorhinus imperata* of 20 days. Breeding may continue throughout the year, if the relative humidity of the air is sufficiently high (Goormans and Pujol, 1955).

Initial attack on the field is sometimes very high. Goormans and Pujol recorded losses of up to 50 or 70 per cent. Removal of all dropped fruits from a kola plantation reduces the initial attack. After harvesting, the nuts should be thoroughly inspected and all infested nuts removed prior to storage.

Control of kola weevils in stored produce with the aid of a fumigant (methyl bromide) was proposed by Alibert and Mallamaire (1955). Reinfestation may be prevented by spraying of the storage space with DDT or BHC.

8.7.2 Kola Fruitfly

The larvae of the kola fruitfly (*Ceratitis colae* silv.) occur frequently in mature kola fruits. Their burrowing may cause serious losses. The exit holes of the mature larvae apparently furnish entry points for the kola weevils (Annual Report Gold Coast, 1917; Hargreaves, 1937; Pujol, 1957).

8.7.3 Kola Stem Borer

The adult stem borer (*Phosphorus virescens* 01, *P. gabonator* Thomas) attacks young growth, on which it feeds and lays eggs. The larvae bore into the twigs towards the thicker woody parts of the tree. It seems that the adult stem borers cannot withstand a low relative humidity and they die rapidly when the dry season has set in. The life cycle from egg to adult beetle lasts more than seven months. In the drier parts of the kola belt (e.g. at Ibadan) the borer goes through one life cycle per year. In areas where the rainy season lasts longer, several cycles may overlap.

Seriously affected trees should be coppiced and regenerated. The young growth should be protected against insect attack in order to prevent reinfestation. In young or recently regenerated plantations it is possible to inspect the trees regularly for the presence of adult borers, which are easily caught.

8.7.4 Summary of Pests Recorded

Information on pests recorded on kola ıs summarized in Table 26.

Table 26 Summary of pests recorded on kola

Common name	Scientific name	Family	Pest status	Damage done	Control measures
Brown cocoa mirid	*Sahlbergella singularis* (Hagl.)	Miridae	Major	The piercing and sucking of young and old pods cause die-back and abortion of pods	Spray with chemicals Gammalin 20 or Didimac 25
Cocoa mosquitoes	*Helopeltis bergrothi* (Reut.)	Miridae	Minor	Piercing and sucking of new growth	
Green shield bugs	*Nezera viridula* (L.)	Pentatomidae	Minor	Feed on new growth. Cause yellowing and withering of plant	
Black citrus aphids	*Toxoptera aurantii* (Fon.)	Aphididae		Feeding and sucking injury cause leaves to curl and distort new shoot growth	
Defoliators (caterpillar)	1. *Anomis leona* (Sohaus.) 2. *Anaphe venata* (Btle.)	Noctuidae Noctuidae	Minor/major Minor/major	Larvae or caterpillars defoliate young trees and seedlings	
Kola stem borers	1. *Phosphorus virescens* (Olivier) var. 2. *P. jansoni* (Cheo.) 3. *P. gabonator* (Thomas)	Cerambycidae	Minor/major	Extensive injury to new growth. Causes stag headed plant or death of young plant. Larvae bore through the chupons having a series of exit holes for discharge of plant gum	Chemical control has only been on the trial levels. The larvae could be controlled by using 0.075% dimethogate Rogor as a foliar spray. Carbaryl (Sevin) could be used to spray the adult

(continued)

Table 26 (*continued*)

Common name	Scientific name	Family	Pest status	Damage done	Control measures
Kola pod borers	1. *Balanogastris kolae* (Desbr.) 2. *Sophrorhinus imperatus* (Faust.)	Curculionidae	Major	Larvae and adults feed in the nuts, and on pods on the trees	Phostoxin tablet or Trogocide capsule as fumigant. Soaking of seeds with husks in Gammalin solution
Termites wood eater	*Nasutitermes* sp., etc.	Termitidae	Major	Termites tunnel into roots and stem of trees of any age, resulting in the destruction of the tree	Spray with Aldrin or Diedrin solution
Pod husk miners	*Characoma stictigrapta* (Hmps.)	Noctuidae	Major	Larvae burrow into the pods but do not directly affect the nuts	Spray with Gammalin 20
Scale insect	*Stictococcus sjostedti* (Cku.)	Stictococcidae	Minor	Piercing and sucking fruit stalks and developing pods. Their injury causes stunted growth and pod drop	Spray with Gammalin 20
Mealybugs	*Planococcus njalensis* (Laing)	Psuedococcidae	Minor	Piercing and sucking injury causes yellowing of plants and reduction in yield	Spray with Gammalin 20
Red banded thrips	*Selenothrips rubrocinctus*	Thrypidae	Minor	The feeding injury of the insects causes leaves to wilt and drop, development is impaired and fruits turn brown from deposits of excreta	Spray with Didimac 25 or Rogor 40

Variegated locust	*Zonocerus variegatus*	Acrididae	Minor	Nymphs and adults devour seedling and plants and reduce yield of older plants	Spray with Gammalin or Unden and hand pick insects
Kola fruit	*Ceratitis kolae*	Trypetidae	Minor	Live on the plant part of fruit	—
Flower insects	*Torma kolae*	Miridae	Probable pollinator	Live in flowers	— —
Taylor ant	*Oecophylla longinoda*	Formicidae	Minor	Knit leaves of plant to make tents	Spray with Rogor, or Sumithion
White flies	*Mesohomotema tessmanni*	Psyllidae	Minor	Suck solute from young leaves and flowers	Spray with Rogor
Black citrus Aphids	*Toxoptera curanti*	Aphididae	Minor	Suck solute from young shoot and flowers	Spray with Rogor
Leaf roller	*Sylepta semilugens*	Pyrallidae	Minor/major	Roll, and leaves defoliate	Gammalin or Rogor

8.8 HARVESTING AND PROCESSING

8.8.1 Time and Method of Harvesting

In harvesting, the farmer must allow his fruits to attain the right degree of maturity. These fruits should also be free from pest attack. For this reason, the farm should be kept in a state of good sanitation by removing and destroying fallen fruits, which are often infested by fruitfly larvae. During the harvest any infested pods should be eliminated. A great deal of wastage which occurs in a farmer's kola arises from either harvesting pods before they reach maturity, or from poor handling after harvesting. Delayed harvesting often results in infestation by kola weevils which gives rise to defective nuts with little or no market value.

The immature kola pods are deep green in colour and turn to an inconspicuous pale green or light brown colour when mature. At this stage, before the pods begin to dehisce and drop, they are ready for harvesting, using either a sharp cutlass or a hooked knife tied to the top of a long pole. The pods are either gathered in a heap under the trees from where they were harvested or removed to a central point where they are carefully broken and the seeds extracted. Delay in processing of the heaped pods will enhance weevil damage. This must be avoided.

The removal of the skin or testa in which each nut is enclosed is the next important stage in kolanut processing. The standard preskinning practice in West Africa is to allow fermentation of the seed coats with the nuts in baskets or heaped on bare ground, with occasional moistening. Nuts are sometimes buried in moist sand, in order to soften the testa by partial decomposition. Whichever method is employed the principle is the same. After the testa has partly decomposed the nuts are soaked in water for about eight hours after which the seed coat can be easily removed (Quarcoo, 1965).

8.8.2 Processing

8.8.2.1 On-the-farm processing procedures

The skinned nuts are washed in water and are collected in baskets through which excess water drains off. They are then spread thinly on a mat under shade for about two to three hours. Defective nuts should be picked out at this stage. The good nuts are subsequently placed in unlined baskets and are covered lightly with banana leaves for about five days. This is the curing process. Considerable 'sweating' which reduces the moisture contents of the nuts occurs during this process. After a week they are transferred into baskets, lined with green leaves. The leaves often used for lining baskets are those of *Mitragna ciliata*, *Thaumatococcus*, *Clinogyne*, *Dorax*, *Sarcophrynium* and *Marantochloa*. Banana leaves are occasionally used, but they should be partially dried before use or they may deteriorate rapidly and require frequent renewal.

During the first few weeks of storage, the nuts should be periodically checked to see that they are not heating up. If heating is observed, the nuts should be stirred thoroughly and left uncovered for a few days; if, on the other hand, there is a tendency for excessive drying, the lining of the basket with leaves should be increased and the top of the basket weighted down to check this excessive loss of moisture. Kolanuts may be preserved in good state inside baskets by this traditional method for about 12 months if processing and handling are carried out with care.

Nuts meant for export are sun dried after separating the cotyledons. In order to further hasten drying, the cotyledons may be cut into slices. If this is done, the slices are soaked in water for two to three hours before drying. The sliced nuts are then dried in the sun for several days after which defective nuts and other foreign matter are removed. A few hundred tonnes of such dried nuts are exported annually to Europe and America where they are used for pharmaceuticals and for beverages. Dried kolanuts are also used locally in the dyeing industry.

8.8.2.2 Nut quality

One important factor which adversely affects the quality of the kolanuts is the presence of weevils. Therefore, during the first few weeks the nuts should be carefully and repeatedly examined for weevil infestation. All infested nuts should be removed and destroyed. When nuts are free from weevils, it is necessary to examine them only once a month, and at each examination, the leaves used for lining the baskets should be replaced with fresh material.

Other factors which determine the value of kolanuts are size, flavour, keeping quality and colour. White-coloured nuts are regarded as superior to pink and purple nuts in flavour. However, flavour and keeping quality are determined by the efficiency of curing and storage.

8.8.2.3 Grading, packing, storage

After processing, the nuts may be stored to await a favourable market or sold immediately. They are graded according to size and colour and are packed in special containers by experienced labourers.

The containers which are woven baskets are carefully lined with leaves. The whole bundle is then carefully roped until the final pack is made neat and strong. Each bundle may contain as many as 5000 to 8000 nuts. The bundle may then be kept in storage or railed immediately to the kola market where they are sold wholesale to small retailers.

The traditional method of storage described above is very laborious and time consuming. It is also ineffective in the control of kola weevils. Recently, the fumigation of nuts with Phostoxin has been shown to achieve control of weevils. However, this chemical requires careful handling and use. New storage methods using polythene bags are being investigated.

The changes in dry-matter content of kolanuts (variety Agege Red) were studied from the time that the nuts had been processed and made ready for storage, a few days after the harvest (van Eijnatten, 1966a). At this moment the nuts had a dry-matter content of 41.5 per cent.

According to customary practice, the nuts were stored in baskets lined with fresh banana leaves. From the stored nuts a sample was removed at weekly intervals for the determination of the dry-matter content. Because the initial total weight of the stored nuts was known, the expected total weight of the stored nuts could be calculated with the percentage of dry matter. These figures were expressed as percentages of the initial total weight of the nuts in Figure 30.

During the first nine weeks, the weight decreased gradually with approximately 0.5 per cent of the initial total weight, per week. From the tenth to the fifteenth week the decrease accelerated and amounted to 3 per cent per week. From then onwards the weight fell rapidly, because the remaining quantity of nuts had become too small for efficient storage and the nuts started to dry out and die.

The use of polythene containers is only recommended after the initial period, probably lasting some two weeks during which active processes take place within the nuts. The same is true for the proposal to seal kolanuts into polythene containers, in which any insect pests are killed off, because the nuts cause a rapid rise in the carbon dioxide content of the atmosphere within the containers (Goormans and Pujol, 1955).

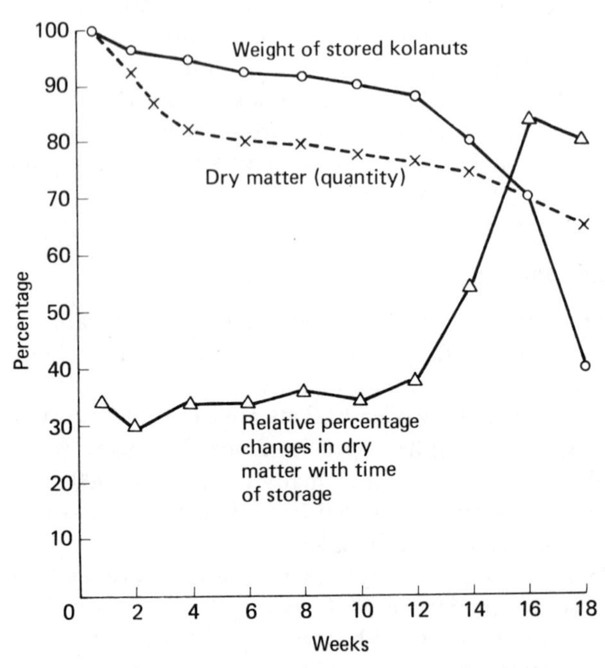

Figure 30. Graph of weight changes in stored kolanuts

8.8.3 Yield Levels

Seedling trees are unlikely to produce any flowers or fruits during the first two or three years from planting, because in this period they still follow their juvenile, monopodial pattern of growth. Until the sixth or seventh year flowers and fruits are scarcely found, although well-developing trees may have fruits before this time. But even afterwards, the trees often bear little. Full productivity is said to be reached between the eleventh and fifteenth years.

Actual observations in various experimental plantations in West Africa showed yield levels from 114 to 588 kg per hectare for trees of 11 to 15 years old, with one outstanding figure of 1893 kg per ha for a plantation at Kumasi, Ghana. Even the highest recorded yield is evidently low, when expressed in the approximate number of fruits. With the figures on percentage weight of kolanuts per follicle and the average follicle weight given by Dublin (1965), the productivity in this case can be estimated at 24 fruits per tree.

8.8.3.1 Improving yields by the use of ramets

The possibility of raising vegetatively propagated plants makes it feasible to establish kola plantations with clonal materials, derived from trees with high yields. The development of such materials and the method of establishing them were discussed earlier.

One of the important characteristics of ramets, raised from cuttings, is their continuation of the growth rhythm of mature trees and their consequent ability to produce flowers and fruits from the very first year. However, it seems unwise to allow the formation of fruits before the end of the second year after transplanting. The ramets need this period to become well established and to form a good root system and canopy.

Fruit production in the first or second year results in nuts which are of a small size and barely marketable. The nuts produced on two-year-old ramets at Agege, Nigeria, weighed only 5.0 to 7.7 g; in the third year the average nut weight varied from 11.7 to 16.3 g and these nuts had become marketable. When the ramets are well established, they can be allowed to bear fruits in their third year.

How the production per hectare from young kola ramets can be further increased by planting at very close spacing, has been described earlier.

8.8.3.2 Improving yields by controlled pollination

The low efficiency of natural pollination is one of the factors limiting the productivity of kola trees. Fruit setting can be improved manyfold by assisting the transfer of compatible pollen to hermaphrodite flowers. Picking a freshly opened male flower from a compatible parent and the actual transfer of the pollen grains to the stigmas of the hermaphrodite flowers to be pollinated take a matter of minutes, provided the pollinator tree is close by. Usually, compatible pollen can be obtained from any neighbouring seedling tree, or in the case of clonal plantings from the alternative clone, when two compatible clones have been planted in the same plantation.

In this way an increase in the proportion of successfully pollinated flowers can be achieved easily, provided the hermaphrodite flowers are readily accessible. Especially in young ramets all hermaphrodite flowers will be within easy reach during the first few years. In seedling trees and in older ramets the use of a stepladder is necessary to effect pollination. The accessibility of hermaphrodite flowers is one of the important reasons why selection of kola trees and their management should be directed towards obtaining lower trees. Pruning methods are to be developed to keep the trees below a height of 7 to 9 m.

8.8.3.3 Improving yields by interrupting normal growth

Wherever kola trees are grown, mutilated or wounded trunks are a common sight. It is thought that wounding the trunk with a machete before the main flowering season induces the trees to produce higher yields during the subsequent season. The usual procedure is to slash the trunk with a machete all round. The cuts do not only mutilate the bark, but they penetrate into the wood.

More sophisticated methods have also been used. A common practice in Guinea is the removal of the bark of four- or five-year-old trees at 10 cm above ground level, in a ring with a width of several centimetres. Partial ringbarking of the tree by removing a 3 cm wide strip of bark at one side of the tree or by removing a strip of bark in the form of a spiral with the turns 30 cm apart was reported to increase yields effectively (Drugs-kola, 1933).

In only one case has the effect of wounding the trees been investigated (Russell, 1955): 40 trees were slashed and compared with 130 trees, which had not been treated. Two years' yield records of these trees were available prior to the initiation of the experiment. Of the slashed trees 54 per cent yielded more in the two years after the treatment than in the previous two years. This proportion was 30 per cent for the untreated trees; the difference being significant at $p = 0.01$.

Bodard (1962) suggested that the change in the nutrient balance of the plant resulting from the interruption of the normal distribution of photosynthates in the plants, caused by the wound in the trunk, decreases the number of buds remaining dormant. He made this plausible by ringbarking individual flushes, a proportion of which produced more inflorescences than usual. The increased number of male and hermaphrodite flowers would then increase the probability of some of the flowers setting fruit.

8.8.3.4 Selection of new kola varieties

Present information points to several criteria which have to be applied to the selection of new kola material, apart from its productivity. The most important ones are as follows:

1. *Compatibility* It is desirable that selected clones are self-compatible. It is obligatory that they show a high compatibility towards other selected clones.

2. *Ability to be propagated vegetatively* The success percentage in rooting of cuttings varies considerably from one seedling tree to another. With the present methods some trees have a success rate of up to 80 or 90 per cent, others as low as 10 per cent.

3. *Ability to transfer a large nut size to progeny* Considerable differences have been observed in this respect and the influence of the female parents is to be noted especially.

4. *Tree size* Seedling trees vary greatly in size. This character is usually retained in vegetative progeny. Smaller trees are easier to maintain and are more accessible for artificial pollination and for harvesting.

5. *Distribution of the yield over the year* Most kola trees produce fruits in two or three particular months of the year, varying with the area. However, some trees have a high percentage of 'out of season' fruits. These are of great importance in West Africa, because prices fluctuate considerably with seasons.

Intercrossing selected clones may lead to combinations of various desirable characteristics. It is the only way for the establishment of new, improved material for further selection purposes and for a study of the heritability of various desirable characteristics in the parental clones.

8.9 KOLA TRADE

There are three distinct stages in the kola trade. These are the trade in unprocessed wet nuts, the bulk sale of processed nuts and the retail trade in both unprocessed and processed nuts. *Cola nitida* is the kola of commerce and it features prominently in the three levels of trade in kola. *C. acuminata* is only of local importance especially among the Yoruba tribe of West Africa and the three stages of the kola trade.

Unprocessed nuts are sold in two forms. The nuts can be sold in the intact pods. This practice is very common among farmers with small quantities of kola. It is also very common with *C. acuminata*. To the inexperienced, it is a risky business to buy nuts in the intact pod, but with the experienced kola trader, such a practice is usually considered advantageous because it affords an opportunity for buying nuts at the cheapest price. Nuts are also sold after they have been extracted from the pods but before skinning. Unskinned nuts sell for a much higher price than nuts in the pods. When large quantities of nuts are to be sold unskinned, they should be graded according to size. Most farmers sell their nuts unskinned. Kola traders also prefer to buy unskinned nuts for bulk processing and storage.

The main function of the primary kolanut buyers, apart from those who may buy a few nuts for personal consumption, is processing and storage. After processing and storage (period indefinite depending on price movements in the kola market) the primary kola buyers sell their graded nuts in bulk and in multiples of 100 to the secondary kola buyers. The secondary kola buyers are itinerant kola traders.

At the processed kola market, the nuts are sold in bulk to the kola retailers or the kola exporters. The kola retailers, and there are hundreds of them, buy quantities which are within their capacity to dispose of within one to one and a half months after which they replenish their stock by buying more.

Kola exporters deal in fresh nuts only for sale to countries within or neighbouring West Africa. Overseas markets demand dried kolanuts.

8.10 USES OF KOLA

Kola has been used in many ways. These various ways can be grouped into industrial and local or traditional uses:

Industrially the kolanut is used:

1. for the preparation of kola-type beverages, such as Coca Cola, Pepsi Cola, kola wine;
2. in the preparation of choca-cola: a type of chocolate containing cacao and kola powder in cocoa butterfat;
3. as a source of alkaloids (caffeine and theobromine) in pharmaceutical preparations.

Kolanuts are traditionally used as masticatory agents for their stimulating effects. It has been claimed by kolanut consumers that kolanuts suppress hunger, thirst and sleep. It is a common custom among long distance drivers in Nigeria to chew considerable quantities of kolanuts en route. It is also said that kolanuts strengthen dental gums and suppress gout and related diseases. Some of the nuts are used as a source of dye.

Other parts of the kola tree have also been put to a number of uses:

1. the pod husk, when mixed with some ingredients, is used in traditional concoctions to reduce pain;
2. the bark is used in treating swellings and fresh wounds;
3. the roots are excellent chewing sticks. The roots are effective in cleaning the teeth and supposedly in strengthening the dental gums;
4. the wood is used in carvings, coach work and boat building. The tensile strength of the xylem fibre is very high.

8.11 FUTURE OUTLOOK ON THE KOLA INDUSTRY

From the little that is currently known about kolanuts, it appears that more research on their uses could yield very rewarding dividends. As in the case of cacao, only the cotyledons of kola have been put into economic use while the pod husk and the fleshy, sugary testa have hitherto been wasted. Research is needed on how best to employ these potentially valuable but hitherto wasted products of the kola industry.

Kola improvement through scientific research began to receive active attention with effect from the 1940s. As a tree crop, research data on kola are limited. There is need to collect more data on production, productivity of the different varieties and clones under the various ecological zones in which kola is grown in West Africa: there is an urgent need for intensive selection for productivity, compatibility and desirable nut characteristics. Other crops (food crops and commodity crops) will certainly compete with kola for resources, but kola appears to have an edge on cacao and possibly coffee because the nuts are widely consumed in West Africa while at the same time kola is a minor export crop.

ADDITIONAL READING

Alibert, H. and Mallamaire, A. (1955). Les charançons de la noix de cola en Afrique; Moyens de les combattre. *Bull, Peot. Vég. Gouv. Gen. Afe. Occ. Franç. Dir. Gen. Serv. Econ. Imp. Gen. Agric.*

Annual Report (*1917*). Agric. Dept. Gold Coast.

Bernegan, D. L. (1908). Studien uber die Kolanuss. *Trgren pfl.*, **12**, 117–126.

Bodard, M., (1955). Contributions à l'étude de *Cola nitida*, Croissance et biologie florale, *Centre Rech. Agron. Bingerville, Bull.*, **11**, 3–28.

Bodard, M., (1962). Contribution à l'étude systematique du genre cola en Afrique Occidentale. *Annales Fac, Sci. University du Dakar.*

Boelman, H. A. C. (1940). De mogelykheelen van de cultuur van einige minder bekende geneerkruiden en huis toepassing. *Bergcultures*, **14**, 1160–1166.

Bontekoe, M. A. (1950). De kola noot. *Tydichr. Econ. Soc. Geogr.*, **41**, 115–120.

Busson, F., Garnier, P. and Dubois, H. (1957). Contribution à l'étude chinnique des cotyledons de *Cola nitida. J. Agric. Trop. Bot. Appl.,* **4**, 657–660.

Caswell, G. H. (1962). *Agricultural Entomology in the Tropics.* London.

Chevalier, A. and Perrot, E. (1911). Les végétaux utiles de l'Afrique tropicale française. fasc. VI. Les kolatiers et les noix de kola. Paris.

Clay, D. W. T. (1963). Vegetativa propagation of kola. *Trop. Agric. Trinidad*, **41**, 61–68.

Cocoa Research Institute of Nigeria. Advisory Leaflet No. 8.

Drugs Kola (1933). *Bull. Imp. Inst.*, p. 31.

Dublin, P. (1961). Colatiers, *Rapp. Ann. 1961, Centre Rech. Agron. Boukoko. Rep. Centreafricaine*, 213.

Dublin, P. (1965). Le colatier en Republique Centreafricaine, Culture et Amelioration. *Cafe, Cacao, Thé*, **9**, 97–115, 175–191, 294–306.

Farmers' Field Day (1972). Cocoa Research Institute of Nigeria, Ibadan, Nigeria.

Golding, F. D. (1937). Further notes on the food plants of Nigerian insects IV. *Bull. Ent. Res.*, **28**, 5–9.

Goormans, C. and Pujol, R. (1955). Rechercher sur la charançon de kola. *J. Agric. Trop. Bot. Appl.*, **2**, 263–280.

Hargreaves, E. (1937). Some insects and their food plants in Sierra Leone. *Bull. Ent. Res.*, **28**, 505–523.

Hutchinson, J. and Dalziel, J. M. (1958). *Flora of West Africa*, Revised by R. W. J. Kear, London.

Kolanuts (1959). *Trop. Sci.*, **1**., 112–114.
Kolanuts and kola chocolate (1953). *Int. Choc. Rev.*, **8**, 185–186.
Mallamaire, M. A. (1954). Catalogue des principaux insectes, nematodes, myriapodes et accarens nuisibles aux plantes cultivées en Afrique Occidentales Francaise et an Togo. *Cong. Prof. Veg. Prod. sons les climats chands.* Inst. Franc. d'Outremer, Marsailles.
Mascre, M. and Paris, R. (1946). Fruits et grains d'outremer utilisés en thérapeutique. *Fruits d'outremer*, **8**, 226–230.
Miles, A. C. (1930). Cola survey of eastern Ashanti area and a general review of the cola industry. *Dept. Agric. Gold Coast, Yearbook 1930, Bull.* **23**.
Nzekwu, O. (1961). Kolanut. *Nigerian Magazine*, **71**, 298–305.
Odegbaro, O. A. (1973). Regeneration of old kola trees: *Cola nitida* (vent) Schott and Endlicher by coppicing. *Turrialba*, **23**, 334–340.
Ogutuga, D. B. A. and Daramola, M. A. (1976). Changes in the chemical composition of kola nuts (*C. nitida*) during storage. *Nigerian Agricultural Journal*, **13**, 124–129.
Onochie, C. F. A. and Stanfield, D. F. (1960). *Nigerian Trees.* Government Printer, Lagos.
Paula, R. D de G. (1938). A NOZ de kola no Brasil, *Inst. Nac. Techn.,* No. 11, Rio de Janeiro.
Progress in Tree Crop Research in Nigeria (1971). A commemorative Book: Cocoa Research Institute of Nigeria, Ibadan, Nigeria.
Pujol, R. (1957). Etude preliminaire des principaux insecter invisibles aux colatiers, *J. Agric. Trop. Bot. Appl.*, **11**, 241–264.
Pujol, R. (1962). Charançons invisibles aux noix de cola. *Café, Cacao, Thé*, **6**, 105–114.
Quarcoo, T. (1965). Development of kola and its future in Nigeria. *Proc. Agric. Soc. Nigeria*, **6**.
Quarcoo, T. (1975). *A Handbook on Kola.* Cocoa Research Institute of Nigeria, Ibadan, Nigeria.
Russell, T. A. (1955). The kola of Nigeria and the Cameroons. *Trop. Agric.*, **32**, 210–240.
Schumann, K. (1900). Uber die Stammpflauzen der kolanusz. *Tropenpfl.*, **4**, 219–223.
Some recent work on insect pests in Sierra Leone (1927). *Bull. Ings. Hist.* **25**, 303.
Sprecher von Bernegg (1934). *Tropische und subtropische Welturrtschaftspflanzen.* III. *Genuszpflanzen.* Vol. 1. *Kakao und Kola*, pp. 214–256. Stuttgart.
Swarbrick, J. (1964). A note on rooting of kola cuttings. *J. Exp. Agric.*, **32**, 225–227.
Todd, J. M. (1951). An indigenous source of swollen shoot disease of cacao. *Trop. Agric. Trinided*, **6**, 109–126.
Van Eijnatten, C. L. M. (1964a). The development of the kola tree and its produce. Cocoa Research Institute of Nigeria, Memorandum No. 3.
Van Eijnatten, C. L. M. (1964b). Studies on the germination of kolanuts. Cocoa Research Institute of Nigeria, Memorandum No. 4.
Van Eijnatten, C. L. M. (1965). *Annual Report 1964/65, Cocoa Research Inst. Nigeria.*
Van Eijnatten, C. L. M. (1966). Some notes on the taxonomy of cola; Cocoa Research Institute of Nigeria, Memorandum No. 7.
Van Eijnatten, C. L. M. (1966a). Effects of different densities of shade on young Kola plants. *Nigerian Agric. J.*, **3**, 31–34.
Van Eijnatten, C. L. M. (1966b). Kola. *Annual Report 1965/66, Cocoa Research Inst. Nigeria.* Ibadan.
Van Eijnatten, C. L. M. (1967). Kola Pollination 1965. *CRIN Memo.* No. *11.*
Van Eijnatten, C. L. M. (1969). *Kola: its botany and cultivation.* Royal Tropical Institute, Amsterdam, Netherlands.
Voelcker, O. J. (1935). Cotyledon colour in kola. *Trop. Agric. Trinidad*, **12**, 231–234.
Warburg, O. (1902). Die Togo kolanüsse. *Tropenpfl.*, **6**, 626–631.
Wills, J. B. (1962). *Agriculture and Land Use in Ghana*, London.

CHAPTER 9

Coffee

The cultivated *Coffea* species have been domesticated in Africa. The Harrar tribe was the first to cultivate *Coffea arabica* in Ethiopia, which is the centre of origin of this species. The use of arabica coffee seeds dates back many centuries. Both the Bible and the Koran make references to the use of coffee beans both for consumption and as a means of exchange. In AD575 Felix of Arabia introduced coffee to the land of the Levants. The spread of the cultivation of *Coffea arabica* has been traced through Yemen to Bourbon and Mauritius, from where it was introduced into the Mascarene Islands, islands of the Indian Ocean, South China, Indo-China, South America.

The centre of origin of the robusta type of coffee is less accurately known. Robusta coffee is the produce from *C. canephora*, which is mainly spread throughout the equatorial zone of Africa from Guinea through Zaire into Uganda. This wide dispersal is an indication that more work is needed to identify the centre of variability of the robusta coffees.

Another, less important, African species, *C. liberica*, is cultivated in various West African countries.

Coffee drinking has become popular among all the nations of the world. Coffee production has thus become an important agricultural activity in the different parts of the tropics. Many types of improved coffee material were introduced into West Africa for experimental plantings. Records of these are, however, not available.

Information on export of coffee from West Africa dates back to 1896 when 15.5 tonnes were exported from Nigeria followed by 25.5 tonnes in 1901 and 32.2 tonnes in 1909. Although coffee production has increased over the years in West Africa, especially in Ivory Coast, the total production has remained very low (Table 27).

Table 27. *Coffea* bean production in major *Coffea* growing countries of Africa (thousand bags of 60 kg each)

	Year					
Countries	1960	1961	1962	1963	1964	1965
Ethiopia	1615.00	1687.00	2124.00	2212.00	2320.00	2907.00
Guinea	239.50	252.75	198.82	175.60	175.00	—
Sierra Leone	86.37	84.67	40.67	32.49	85.23	69.04
Ivory Coast	2210.00	3185.00	1536.00	3312.00	4352.00	3363.00
Ghana	39.83	27.90	63.95	44.23	46.23	56.00
Togo	61.07	160.25	192.37	118.00	310.00	194.00
Republic of Benin	13.92	34.83	28.82	16.70	45.52	—
Nigeria	78.73	*10.17	13.40	26.30	76.21	84.32
Cameroon	527.00	749.00	638.00	750.00	858.00	912.00
Rio Muni and Fernando Po	155.00	146.70	123.30	117.00	132.12	137.05
Gabon	10.00	15.00	17.00	19.00	18.00	19.00
Congo Brazzaville	4.98	11.30	17.82	10.70	12.13	11.85
Central African Republic	98.70	121.00	131.40	96.30	130.60	118.00
Zaire	878.00	499.30	551.50	787.90	792.00	814.25
Burundi and Ruanda	381.00	410.00	375.00	188.00	421.00	396.00
Angola	1610.00	2651.00	2784.00	3035.00	2765.00	3283.00
Uganda	1880.00	1812.00	1947.00	2998.00	2900.00	2095.00
Kenya	397.00	555.00	550.00	558.00	718.00	646.20
Tanzania	422.90	484.50	298.50	463.50	604.00	560.00
Madagascar	583.00	917.00	767.00	900.00	833.00	874.00

Source: Krug, C. A. and De Poerck, R. A. (1968) World Coffee survey, F. A. O. Agric. studies.
* Western Cameroon which hitherto produced the bulk of Nigerian Coffee opted to join the Federal Republic of the Cameroon.

9.1 TAXONOMY

9.1.1 The Genus *Coffea*

Coffea is a member of the family Rubiaceae, a large family of over 5500 species widely distributed throughout the tropics. The genus *Coffea* is characterized by simple opposite leaves, united calyx, united corolla, an inferior ovary and generally two locules. The genus includes about 100 species, some of which are yet to be properly classified. The plants are evergreen shrubs, thriving very well in the hot wet tropical part of the world. There are three basic coffee types, namely the tetraploid self-pollinating *Coffea arabica* with a somatic chromosome number $2n = 44$. This is highland coffee and it grows very well at altitudes of 600 m and above. *Coffea liberica*, $2n = 22$, is a mid-altitude coffee thriving best at an altitude of 400 to 600 m, and *Coffea canephora*, $2n = 22$, lowland coffee thriving best at the altitude range of 0 to 750 m above sea level.

The genus *Coffea* has been divided into four sections, namely:

1. *Para-coffea* Miquel
2. *Argo-coffea* Pierre ex de Wild
3. *Mascarocoffea* Chevalier
4. *Eucoffea* K. Schuman

The mascarocoffea is characterized by the complete lack or near absence of caffeine in most of its species. The section Eucoffea contains all the important commercial coffee species.

By far the most important species of cultivated coffees are *C. arabica, C. canephora* and *C. liberica* in that order.

9.1.2 *Coffea arabica*

Coffea arabica is the original source of coffee beans and it produces the best quality coffee. Generally, arabica coffee trees are slender. The crop originated from the Ethiopian Highlands. Cultivated coffee trees are usually heavily branched, dimorphic in growth and the trees are invariably kept in shape by judicious pruning.

Leaves are simple, alternate, opposite, thin, dark green, shiny surfaced, fairly stiff; axillary and sub-axillary buds often develop into the reproductive lateral branches. Leaves are petiolate. In some cases, leaves bear interpetiolar stipules. The leaf midrib and the lateral veins are prominent. Flowers are produced in dense clusters along the reproductive branches in the axils of the leaves. Usually, and depending on the suitability of the environment, *C. arabica* trees produce two or even three flushes of flowers per year. Flowers are white, sweet scented, star-shaped and carried on stout but short peduncles. Bracteoles are united forming a cup-shaped epicalyx at the base of the flower. There are five calyx segments which unite into a short tube. The corolla is gamopetalous, five segmented halfway the length, the segments spreading out very widely at anthesis. There are five stamens which are inserted in the corolla tube. The anthers are carried on long, slender and upright filaments. The ovary is inferior, two united unilocular carpels, each containing a single ovule attached to the base of the carpel wall (basal placentation). The ovary bears a slender style which terminates in bifid short and pointed stigmas (see Figure 31).

Coffea arabica is naturally self-pollinating. The fruit is a berry. The pericarp is composed of an outer thin shiny exocarp, a fleshy mesocarp and a relatively thin but tough endocarp in which the seeds are enclosed. *Coffea arabica* fruits take seven to eight months from pollination to maturity. Immature berries are

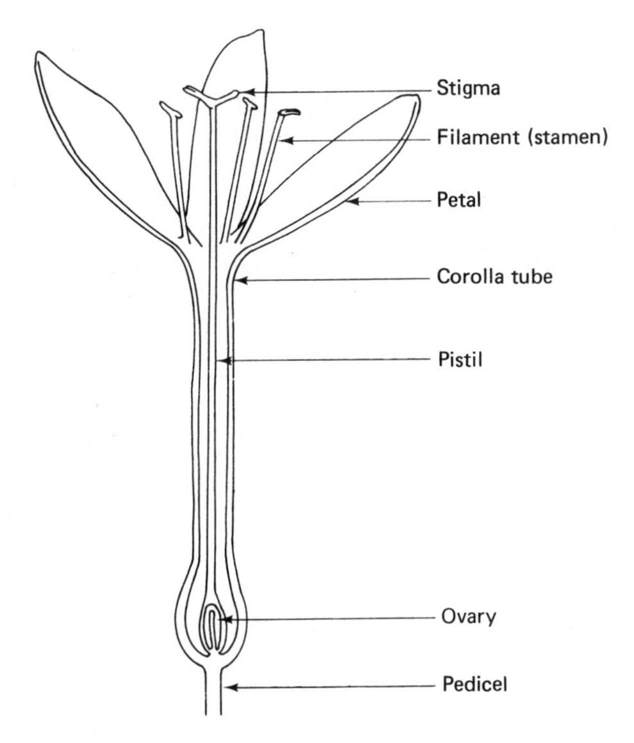

Figure 31. Floral structures of *Coffea*

dull green, and on ripening the skin colour changes through yellow to bright crimson. Each berry contains two seeds, tightly pressed together at their flat surfaces; the inner surface of the seed is deeply grooved. The seed is composed of the testa, the embryo and a curiously folded endosperm.

Plant breeders through the ages have produced many varieties of *Coffea arabica*. These varieties differ in their responses to ecological factors – soil fertility, rainfall, disease and pest reaction, vigour hardiness – and in the quality of their beans. *C. arabica* is generally homozygous. Therefore, intravarietal tree variation is low.

9.1.3 *Coffea canephora* (Robusta Coffee)

The robusta coffee is a lowland coffee with its centre of origin in equatorial Africa. Other robusta types are similar in vegetative and reproductive characteristics and have their centres of origin also in the rain forest areas of tropical West Africa. These types comprise:

Coffea excelsa
Coffea congoensis
Coffea marcrocarpa
Coffea abeokutae
Coffea stenophylla
Coffea brevipe

The quality of the beans of the robusta-type coffees is inferior to that of *C. arabica*, but distinctly superior to that of *C. liberica*. The robustas are lowland coffees, very tolerant of coffee leaf disease (brown leaf disease of coffee) and high yielding under satisfactory environmental conditions.

The flowers are produced in clusters often mixed with leafy bracts around leaf axils. The leaves are thick and larger than those of *C. arabica*; they possess wavy margins and are more or less rounded at the base. *Coffea canephora* trees are much more vigorous than those of *C. arabica*, especially when grown under tropical lowland conditions. The robusta types of coffee are outbreeding.

9.1.4 *Coffea liberica*

This species has its centre of origin in Zaire, although a few authorities have claimed Liberia in West Africa as its centre of origin. *C. liberica* is characterized by very large seeds (about 2000 beans/kg) and the trees are very vigorous, leaves slightly larger than those of *C. canephora* and well adapted to medium altitudes (450 to 600 m above sea level). It is outbreeding.

9.2 AGRONOMY OF COFFEE

9.2.1 Selection of a Site

The selection of sites for planting coffee is determined in the first place by the type of coffee to be grown. There are three groups of coffee, classified on the basis of their requirements for altitudes – the highland coffee, *C. arabica* (600 m and above); the medium altitude coffee, *C. liberica* (450 to 600 m); and the lowland coffee, the robusta-type coffees (0 to 750 m). The type of coffee to be grown therefore determines the altitude at which to site the plantation. In addition to altitude (which is very critical especially for *C. arabica*), the climate (rainfall, temperature, wind) and soil factors exert a considerable influence on the selection of a site for planting coffee.

9.2.1.1 Rainfall

Coffee can grow under a very wide range of rainfall – 750 to 3000 mm per annum. Although coffee is generally more drought tolerant than cacao, the ideal area should have its rainfall distributed evenly over eight to nine months of the year. Coffee is much less sensitive to moisture stress than cacao and kola.

9.2.1.2 Temperature

As a tropical tree crop, coffee (except *C. arabica*) requires high temperatures for good growth and productivity. All coffee types are very sensitive to frost, which leads immediately to rapid leaf and fruit drop. Coffee should be grown in frost-tree areas.

9.2.1.3 Relative humidity

Relative humidity is very important in the cultivation of coffee. Coffee can tolerate high humidity but very low relative humidities of 5 to 10 per cent lead to rapid leaf and fruit shedding. Under very low humidities, coffee plants can remain leafless but alive for a considerable period.

9.2.1.4 Wind

Coffee trees are slender and branches easily break when bearing a heavy crop. Coffee should not be planted in areas subjected to strong winds. Also, coffee does not react favourably to desiccating winds. If coffee has to be planted in a windy area, the plantation should be protected with a windbreak.

9.2.1.5 Soils

Very sandy soils and shallow soils are unsuitable for growing coffee. Hillsides with a gentle slope and deep fertile soils are ideal. Soils for coffee should be deep, slightly acid, well-drained loams. They should be rich in nutrients especially potash and with a generous supply of organic matter. Savanna soils with adequate rainfall, moderate acidity to neutral or slight alkalinity are also suitable for coffee.

9.2.2 Nursery Aspects

9.2.2.1 Raising coffee seedlings

The general practice of sowing coffee seeds with parchment to raise seedlings has generally resulted in failure. Some farmers sow coffee berries or worse still coffee seeds that have been stored for a long time. This results in irregular germination and production of non-uniform seedlings.

One problem with coffee seed is to keep it alive for a considerable time. The live seed of coffee has a hard parchment. Seeds should be enclosed in the intact hard endocarp or parchment for storage. This is so because removal of the endocarp exposes the delicate, silvery and fragile testa or the true seed skin, which surrounds the endosperm within which the embryo is enclosed. Another problem related to seed viability is that both the endocarp and the testa are hygroscopic. This makes the storage of coffee seeds difficult. The problem of storing coffee seeds for a long period under viable conditions needs research, as it is known that the longer coffee seeds are kept in storage after harvesting, the further the percentage germination falls. Such seeds also give variable and poorly developed seedlings.

9.2.2.1a Selection of seeds

Coffee seeds intended for raising seedlings should, therefore, be selected before sowing. Selection of coffee berries for germination aims at retaining seeds that are viable and healthy, and which will germinate rapidly.

Pea-berry seeds as well as those damaged as a result of insect and/or fungal attack are not likely to be viable. Plump, clean seeds are likely to be viable. The first step is to separate seeds into the above groups. A rough test and rule of thumb is to put the seed lot in water. Pea-berry and damaged seeds will float, but healthy and potentially viable seeds will sink. If any of the potentially viable seeds shows noticeable features which suggest that its seedling will be abnormal, it should also be eliminated at this point.

9.2.2.1b Sowing

When seeds with the parchment are sown, the endocarp prevents the free movement of gases, particularly of oxygen, to the embryo. This limits respiratory activity and subsequently the growth of the seedlings. In order to minimize seed losses during germination and also to ensure healthy, vigorous seedlings, seeds should therefore be hulled and dressed with a fungicide.

The selected seeds are sown in seed boxes or trays at 2.5 × 2.5 cm in a medium of fine river sand. The sown seeds are watered and the boxes covered with wet sacking or transparent polythene sheeting in order to keep the medium moist. Waterlogging should be avoided as at this stage it could increase pre-emergence losses.

Germination takes place in four to six weeks, depending on prevailing temperatures.

9.2.2.1c Potting of the seedlings

Seedlings should only be transplanted after the cotyledons have opened and before the seedlings pass the four leaf stage.

The seedlings should be lifted from the germination box to polythene pots (25 × 12 cm) already filled with topsoil or to the prepared seedbed as the case may be. The potting mixture may be either of rich surface soil or one part topsoil and one part compost.

Before lifting the seedling, it should be watered. Two fingers are used to dig around the roots until the entire root system is exposed. A hole a little more than the size of the roots should be made in the soil at the centre of the potting medium. Seedlings should then be carefully lifted from the tray and the taproot placed in the hole at the centre of the potting medium. The soil should be carefully pressed around the roots. Watering should be done immediately after potting. Subsequent watering should be carried out as often as necessary until the seedlings are ready for transplanting.

It is important to avoid excessive moisture because it leads to waterlogging. In such cases root rots may kill the young seedlings. If waterlogging is observed, watering should be suspended.

9.2.2.1d Shade

Seedlings should be raised under light overhead shade (50 per cent shade). Starting from six weeks to transplanting the shade should be gradually removed.

Seedlings will be big enough for transplanting to the field in about five to six months.

9.2.2.1e Summary of activities

Where trays or seed boxes are not available, the shelled seeds could be sown on seedbeds at a spacing of about 2.5 × 2.5 cm.

1. Put freshly harvested cherries in water and discard floaters.
2. De-pulp the selected ripe cherries (adopting the normal processing method).
3. Seeds intended for germination should not be fermented.
4. Wash thoroughly to remove mucilage.
5. Spread under shade to air dry the seed.
6. Shelling of the dried, de-pulped berries or parchment should be carried out particularly when they are to be sown under warm humid conditions.
7. Use manual shelling to avoid mechanical damage.
8. Apply seed dressing to the selected seeds by soaking in a fungicidal suspension (Thiram or Captan) for 24 hours and then sow in seed trays.
9. When the radicles have emerged, the seedling should be transferred into a seedbox.
10. Water as often as necessary.
11. Seedlings should be potted a few days after the first foliage leaves have unfolded (six to eight weeks from sowing).

9.2.2.2 Vegetative propagation

Although coffee is usually propagated by seed, it has been successfully propagated by rooting, layering, marcotting and budding. Vegetative propagation has been used very widely in coffee breeding especially when clonal materials are to be tested on a wide scale. For rooting of coffee cuttings the single leaf-bud cutting is commonly used. It has been claimed that this has been more successful with coffee than the multiple leaf-bud cutting.

9.2.3 Planting Operations

9.2.3.1 Land preparation

Coffee is generally transplanted into the field with nurse shade. Modern practice with coffee requires that a temporary shade is provided for coffee seedlings for the first 12 to 18 months after transplanting into the field. After this period, the nurse shade is removed. The land preparation is clear-felling with excess thrash removed or burnt. The field is then divided into blocks which are lined out at the chosen spacing, usually 3 × 3 m. Next, the planting holes are dug at 60 × 60 × 60 cm.

Another method of land preparation which is receiving wide acceptance is the Advance Land Preparation Method. In this method, the land for planting coffee is prepared 12 to 15 months ahead of transplanting. The method consists of:

1. clear-felling;
2. burning or removal of excess thrash;
3. establishment of a legume ground cover.

The seedlings should be transplanted into the prepared field when the rains are steady.

9.2.3.2 Transplanting

Transplanting consists of three distinct operations after planting holes have been dug:

1. Uprooting the seedlings from the nursery beds, selection of plantable seedlings and transportation of the seedlings to the planting site.
2. Placing a seedling at each planting hole; removal of the polypot where necessary, placement of the seedling in the planting hole.
3. Filling the planting hole with topsoil and consolidating the soil around the roots of the seedling. Shade and ground cover are adjusted and where overgrown seedlings have been transplanted, some of the leaves are removed to reduce transpiration.

Coffee seedlings are transported to the field after land preparation has been completed. Care should be taken when transplanting to avoid damage to the seedlings. Once a seedling has been placed in the planting hole the soil should be properly consolidated around it.

Poorly developed seedlings should not be transplanted to the field. Where seedlings are raised in polythene pots less care and labour are needed at transplanting. At transplanting the polythene pots should be torn off. Where seedlings are to be transplanted from the nursery bed to the field, care should be taken when digging out the seedlings to avoid damage to the roots.

In certain circumstances seedlings may need to be transported a distance before transplanting. If they have been raised in pots there is little trouble during transportation provided they are well packed. Where seedlings have to be uprooted before transportation, however, precautions have to be taken to protect the roots. Usually the roots are dipped in a clay slurry, otherwise they should be wrapped in wet sacking.

For better establishment, potted seedlings are normally preferred to seedlings with bare roots.

Stump planting is the method adopted for transplanting overgrown seedlings or mature trees that are growing out of position. It is an expensive method of transplanting and should be avoided as much as possible. In practice, the overgrown seedling or the mature tree is pruned back (dehorned) to stumps of about 60 cm in height. The roots are dug out as carefully as possible, avoiding unnecessary damage. The roots are lightly pruned and treated with clay slurry to avoid drying. The stumps are then transported to the planting site and planted within 12 hours after arrival. The size of the planting hole will depend on the size of the stump.

9.2.4 Maintenance of the Coffee Farm

Coffee farms or plantations are maintained in a number of different ways depending on the amount of resources available to the planter, the price and the type of coffee. Management may be intensive and in this case the orchards are regularly weeded, pruned, harvested and adequately protected from pests and diseases. Extensive may be the extreme of harvesting semi-wild trees.

9.2.4.1 Weeding

Removal of weeds from the coffee orchard in the form of clean-weeding is necessary during the establishment phase of the transplanted seedling. Attention to clean-weeding during the first 18 to 24 months helps to eliminate weed problems in later years and facilitates good establishment of both the coffee seedlings and the ground cover. After the first two years the ground cover should have been fully established and from then the ground cover will require more attention rather than the weeds. The ground cover, as the name implies, is to cover the ground and not the coffee plants. The ground cover should be removed within 50 cm from the base of the plant. The exposed soil around the plant should be heavily mulched and kept clean-weeded. Any weed occurring in the ground cover should be removed.

Where labour is scarce, herbicides could be used around the coffee trees to suppress weeds which may grow in the mulched area. Herbicides should not be used on the ground cover.

9.2.4.2 Mulching

Coffee plants are known to benefit from mulching. Organic mulches (banana thrash, sugar cane leaves, elephant grass, leguminous plants) have been shown to be beneficial to coffee. The quantity of mulch applied will depend on the availability of mulching material and the cost of application. Nevertheless, mulch should be at least 15 cm thick and 45 cm in radius. A gap of about 10 to 15 cm radius around the plant should be left unmulched. This is to reduce the risk of insect (termite) attack on the plants. Where termites are prevalent, the mulch should be sprayed with Rogor 40.

9.2.4.3 Shade

A controversy has continued among coffee experts whether coffee is best grown with or without shade. The arabicas, canephoras, libericas and the other coffees have been grown very successfully without shade. On the other hand, many planters claimed beneficial effects of shade on the productivity of coffee. A number of experimental results from West Africa, Brazil and Kenya have not given a clear-cut verdict on this debate. The problem is further complicated by the paucity of studies on microclimatic requirements of coffee.

In its natural habitat, coffee forms part of the under storey of the tropical forest. The arabica of Ethiopia has been recorded to grow in the midst of other genera such as *Albizzia*, *Celtis*, *Clausenopsis*, *Cordia*, *Ficus*, *Mora*. Most of the coffee in West Africa is produced in the Ivory Coast. The generally accepted practice here is to grow coffee in full sunlight with a thick ground cover of legumes. In Nigeria, arabica coffee is grown without shade on the Mambilla Plateau while the robustas are grown with or without shade.

One observation, however, is worth recording. It has been observed by many workers that coffee plants remain alive longer and produce better orchards in shade rather than in monoculture in full sunlight.

The most common shade trees that have been used for coffee are species of *Inga*, *Acacia*, *Albizzia*, *Leucuena*, *Erythrina* and *Glyricidia*. Which of these species of shade trees will suit an area will depend on experimental results obtained for each of them in that particular locality.

9.2.4.4 Systems of pruning

Pruning has become an important maintenance operation with coffee. It has also been shown that high productivity of coffee is directly dependent on good pruning practices. Coffee pruning is a complicated process, which differs from place to place. All systems of coffee pruning are based on an understanding of the growth and morphology of the plant. Growth of a coffee tree occurs in two directions:

1. Upright orthotropic growth – made up of the vegetative leaders or verticals.
2. Horizontal growth – lateral branches known in coffee culture as plagiotropic growth. These lateral or fruit branches are the branches which bear the crop.

A number of coffee pruning systems have been developed. Coffee pruning has developed into an art in many of the coffee producing countries, as Brazil, Costa Rica, Kenya and Ethiopia. Coffee growing countries of West Africa are yet to develop the art.

The different systems of coffee pruning constitute in a special way a system of continual regeneration or renovation of coffee trees in an orchard. Coffee trees respond very easily to regeneration by coppicing and pruning.

9.2.4.4a De-suckering

This is not a pruning system *per se*. It is the removal of unwanted suckers from the coffee plants. Coffee plants often develop large quantities of suckers which must be removed during the normal maintenance operations of the orchard. De-suckering helps to maintain the required number of vertical stems and, at times, it becomes a method of pruning. In practice, young suckers are pulled out during weeding or harvesting. Developed suckers are cut off with a sharp machete or pruning knife. When the cut surface is large, it should be painted over. De-suckering is particularly necessary during the early years of the coffee bush. Coffee plants growing under adverse weather conditions or those whose growing points have been damaged by pests or disease pathogens often produce large numbers of suckers which must be removed or reduced to maintain the desired number of stems.

9.2.4.4b Topping

This is regarded as the simplest method of coffee pruning. It consists of cutting off the upper part of a developed, well grown, vigorous coffee tree. The effect of topping is to cause the vertical and the laterals to thicken. It also helps to reduce the amount of fruiting to that which can be easily borne by the trees. It also prevents tree breakage as it limits vertical growth. Topping, however, is known to induce heavy tillering in some varieties of Robusta coffee. As the fruits of coffee are borne on the laterals, the laterals are handled in such a way as to induce the development of secondary, tertiary and quarternary laterals. The number of such laterals can be controlled by thinning.

9.2.4.4c Single stem

The system makes use of one upright (vertical) stem. All other stems are removed, leaving usually the most vigorous vertical stem to bear the laterals on which the berries are borne. Coffee trees pruned in this way are known to be light in vegetative growth and less demanding on soil nutrients.

Single stems yield fairly heavily and reasonably early. When in crop the laterals drop into a horizontal position under the weight of the berries and later may even bend down. This phenomenon has given the single stem system the alternative name 'parrot perch system'. This system is widely used in West Africa.

9.2.4.4d Double stem rotation

Also known as the 'leaning' or 'modified single stem' or 'bayonet' system. This system of pruning is used when coffee is planted in fertile soil and the growth is luxuriant. In practice, young trees are allowed to produce the first crop before they are subjected to pruning. After harvesting the first crop the stem is topped to about one-third of its height. Shoots (usually two) develop just below the cut end. If more than two shoots develop, the best two are retained while the others are removed. One of the two bayonet shoots is allowed to continue its upright growth. The other is cut back when it is about 10 to 15 cm long (this is the discarded bayonet). Adventitious shoots (suckers) around the discarded bayonet are encouraged to put out laterals for fruiting. In fruit, the retained bayonet will slightly bend over. Meanwhile the discarded bayonet will initiate vertical growth. One vertical is allowed to grow from it. After harvesting the crop of the first bayonet, it is topped, retaining only the old foilage and fruiting wood on the old stem portion. Now, the second bayonet takes over the fruiting function for the year. The process is repeated year after year. After bearing for four to five years the bayonet becomes aged and it is removed to make way for the development of new bayonets. It is essential to observe bayonets for their ageing reaction and develop new uprights to take over before there is any depression in yeild. Water suckers are common in the bayonet system of pruning and they must be constantly removed. Although complicated in theory, it is a simple system to operate once initiated.

9.2.4.4e Upright multiple stem

This system is also known as 'vertical multiple stem'. It is commonly used with *Coffea arabica.* Newly planted coffee trees are allowed to produce the first crop before they are subjected to the vertical multiple stem system of pruning. Two or three young vertical shoots are encouraged to develop from the base and these are topped when about 1.80 m in height. The strongest, oldest and healthiest branch is allowed to set the crop for the year while others are suppressed by pruning or trimming. The selected vertical usually bears for two years. Then it is cut away immediately after the second harvest. The next strongest vertical is then allowed to produce the crop for the next two years; to ensure success, rotation on heavy yielding trees should be at short intervals, maybe even of one season.

9.2.4.4f Leaning multiple stem system

This is a modification of the vertical multiple stem; and it is recommended for areas with fairly high temperatures with good soil fertility, where the coffee is grown under shade and the rainfall is adequate. In practice, seedlings are topped at about 0.5 m height, the shoots are cared for as in the vertical multiple stem system. New shoots to become verticals are encouraged from the base, about one per year. Terminal growth is unrestricted. During growth and bearing, workers go round the plantation several times to gently pull the tops of fruit bearing verticals outwards and downwards to fit them for harvesting. When sufficiently dropped, the tops of the verticals stop growing. At this stage, it is usual to observe a sudden increase in the production of suckers. These are to be removed. Well managed, the verticals can remain in production for three to five years.

9.2.4.4g Hawaiian system

This is an intensive upright stem rotation pruning system. In practice it makes use of four to six uprights. These uprights are in continual use and they are so managed that half of them are in heavy production each year, the other half are held to light fruiting or held in preparation for fruiting. This intensive system is only applicable to coffee trees that are in excellent condition of growth.

9.2.4.4h Colombian system

Also known as the umbrella system, this system is commonly used for coffee grown under shade and when growing conditions are good. In practice the verticals of the young plants are allowed to grow to about 2 m high and then topped. Growth from the terminal two axils is prevented. Laterals grow horizontally, branching repeatedly, and under the weight of the fruits bend downwards, thus forming a dome-shaped or umbrella-like shape. Fruiting branches remain in production for three to four years before they are replaced. In this system, the fruit-bearing branches are known as the bandolas. This is the common system of pruning in the Ivory Coast.

9.2.4.4i Candelabra system

Also known as the 'Costa Rica system'. The candelabra system of pruning starts from the nursery when seedlings are topped at 30 cm. The topped seedlings are transplanted into the field and topped again at a height of 60 cm; from this stage two equal uprights are allowed to develop. These grow for about 1 m and are topped. By this time four uprights have formed (two in pairs) on a single trunk. The tree will be about 3 m in height. Laterals are encouraged to grow for fruit bearing. The heaviest crops are borne on the outside bandolas. When laterals have been in bearing for three to four years, they are cut back to about 20 to 30 cm. A new sucker is allowed to develop: this is topped at the required distance and the candelabra formation is re-established.

9.2.4.4j Guatemalan system

The 'Agobio' or the bent branch system. When coffee verticals touch the ground, the stem forms roots. This is the basis of the Agobio system of pruning (see Figure 32). In practice, trees are trained from the seedling stage onwards. When seedlings are transplanted into the field, they are allowed to establish. Next they are bent over and held down to 45° or an even more acute angle to the soil. This inhibits top growth but induces shoots to emerge from the base of the seedling. Three to four of the verticals are allowed to grow to about 70 cm in length.

The strongest uprights are then selected and the top of the main stem is cut off, resulting in a multiple stem system. When the multiples have grown sufficiently, they are bent to continue the system. As time goes on, vegetative branches spring out of the arch. The branches grow straight, put out laterals for fruiting, thus adding their weight to the best branch. The oldest tops, bent downward are cut off after the harvest as they are no longer useful. The sequence is continued.

The Agobio system of pruning is popular in Asia and South America.

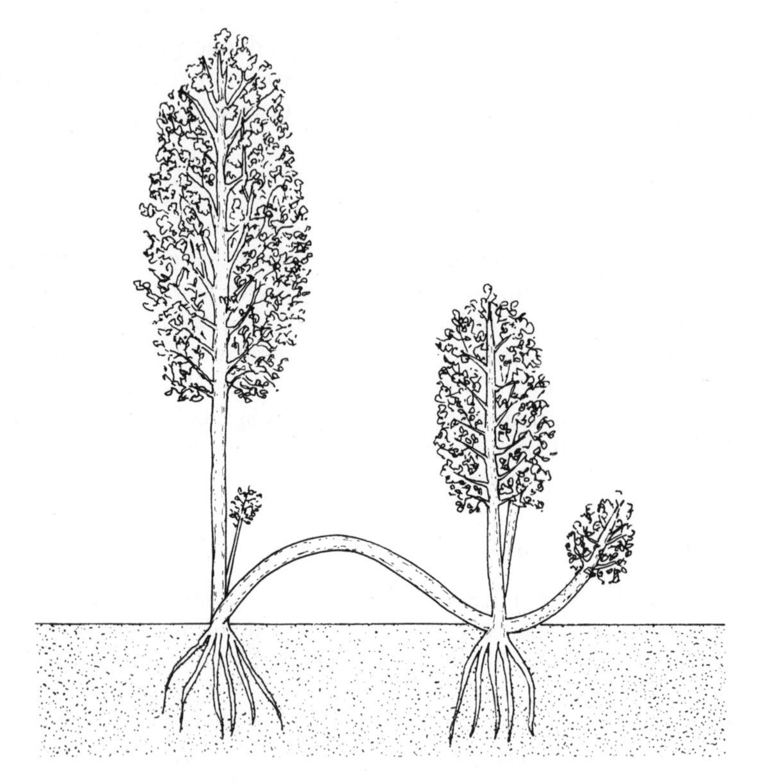

Figure 32. Agobio system of *Coffea* pruning

9.2.4.5 Manures and fertilizers

Experience with the application of organic manures and inorganic fertilizers to coffee in West Africa is limited. This lack of attention to the application of fertilizers to coffee could be due to the fact that coffee is tolerant of low soil fertility and that coffee has not long been grown in West Africa as a commercial crop. The most extensive work on the application of fertilizers to coffee has been carried out at Bingerville, Ivory Coast. This clearly demonstrated that coffee berries contain considerable amounts of nitrogen, phosphorous, potash, iron, zinc, manganese and magnesium. Replacement of these elements should be arranged through fertilization.

It is known that coffee benefits from the application of organic manures either in the form of compost, farmyard manure or composted household refuse. On large plantations, organic manure is made available by decomposing trash, weed, mulch or legume ground cover. When coffee is grown as a compound crop, it is essential that organic manure is supplied in the form of compost. In addition to releasing nutrient elements to the soil, organic manures help to improve the physical fertility of the soil.

Table 28 Major diseases of coffee in West Africa

Name of disease	Causal organism	Plant part	Rating	Symptoms	Control
Hemileia leaf rust (orange rust, common rust or oriential rust)	*Hemileia vastatrix* *H. coffeicola*	Leaves	Major and dangerous	Limited leaf lesions, rounded pustules on the abaxial leaf surfaces, spores flory and orange coloured. Adaxially lesions appear as translucent spots	(a) Use resistant varieties (b) Use copper fungicides (c) Farm hygiene and sanitation
American leaf spot (keleroga; ojo de gallow; gotera)	*Stilbellum flavidum* *Agaricus citricolor*	Foliage and fruits	Major	Lesions on foliage and fruits produce hair-like stalks, the tips of which are slightly flattened	(a) Farm hygiene and sanitation (b) Spray with copper fungicides. Captan can also be used
Black rot	*Pellicularia koleroga*	Foliage and fruits	Major	Dark brown, string-like travel stage, appressing along the undersides of the coffee branches and leaves	(a) Farm sanitation (b) Spray with Bordeaux mixture
Tops disease	*Rhizoctonia* sp.	Foliage and fruits	Minor	Leaves tend to grow more upright than usual when they are attacked	(a) Farm sanitation (b) Use Bordeaux (c) Destroy all diseased material
Aureolate bacterial leaf spot	*Batcerium tumefaciens* *Pseudomonas* sp.	Foliage and fruits	Minor	Leaf defoliation; dark coloured spots with broad yellow haloes on leaves	None available
Brown eye spot	*Cercospora coffeicola*	Foliage and fruits	Minor	Spots (large or small) with a grey centre, few black spots, band of yellow halo; fruit blotches	Use resistant variety

(*Continued*)

Table 28 (*Continued*)

Name of disease	Causal organism	Plant part	Rating	Symptoms	Control
Weak spot	Many fungal species (*Cercospora* sp., *Glomerella* sp. *Pestalozia* sp. *Diplodia* sp.)	Foliage and fruits	Minor	Development of translucent spots, swollen leaf cells, weak spots halo of yellow tissue in a ring	Spray with copper fungicides during dry season
Anthracnose (infectious die-back)	*Pestalozia* sp. *Fusarium* sp. *Glomerella* sp. *Gloeosporium* sp. *Colletotrichum* sp.	Foliage and fruits	Minor	Blackening of coffee leaves, fruits and stems	Use resistant varieties
Blight, burn, quema	*Phoma costaricensis* *Phyllosticta* sp. *Phomopsis* sp. *Ascochyta* sp.	Foliage and fruits	Minor	Dark brown lesions or spots on young foliage, malformation of leaves. Stems and sudden death of leaves and branches	(a) Reduce shade (b) Farm sanitation
Mealy pod	*Trachysphaera fructigena*	Fruits	Severe during wet weather	Dark purplish and brown rot of the fruit tissues. Sometimes rot of foliage occurs	(a) Spray with copper fungicide (b) Farm sanitation
Coffee berry disease (CBD)	*Colletotrichum coffeanum* var. *virulans*	Fruit	Major and popular	Lesions dark bronze to rusty-pink in colour	(a) Farm sanitation (b) Spray with fungicides
Black rot	*Pellicularia kolerage*	Fruit	Minor	Blackening and rotting of berries	Spray with copper fungicide
Grey blotch	*Cercospora coffeicola*	Fruit	Minor	Fruit soft spots	None. Improve soil fertility
Fusariosis	*Fusarium equisetti* var. *intermedium*	Fruit	Minor	Darkening of berries, discoloration and internal blackening of beans	Farm sanitation

Striped bean (zebra)	*Nematospora coryli*	Fruit	Minor	Dark stripes on berries	Spray with insecticide to eliminate vectors (insects)
Fruit blister	*Botrytis cinerea* *Sclerotium coffeicolum*	Fruit	Minor	Blister of berries. Berries become dry, mummified and unsaleable	
Strangulation; stem ulcer	(i) *Rhizoctonia* sp. (ii) *Fusarium* sp.	Stem	Major	Injury and decay of the bark, discoloration of the exposed wood and accumulation of callus tissue around wound lesions. Most common when coffee is grown under frost condition	Avoid exposure of plants to frost
Red disease	*Phytomonas* (*Leptomonas*) sp. *Rhizaecus* sp.	Phloem tissue of the wood	Serious but minor	Sudden wilt and collapse of green trees without their leaves changing to yellow. Leaves remain attached to dead trees	(a) Control the insect vector (*Linus* sp.) (b) Farm sanitation
Tracheomycose (*Fusarium* coffee disease)	(i) *Fusarium xylarioides* (ii) *Carbuncularia xylarioides* (perfect stage)	Stems	Serious but minor	Yellowing and collapse of trees	Identify infected trees very early and remove
Morte subita (soil borne)	*Fusarium* sp.	Vascular (common in Africa)	Minor	Yellowing and rapid Collapse of trees	Use resistant variety
Fusarium stem and root infections	Various species of *Fusarium*	Stem and roots	Minor but could be serious	Common with high altitude coffees	(a) Farm sanitation (b) Avoid growing coffee at too high altitude

Table 28 (*Continued*)

Name of disease	Causal organism	Plant part	Rating	Symptoms	Control
Nectria trunk canker	*Nectria dodgei*	Trunk	Minor	Irregularly shaped, rough bark cankers with small nectria bodies. Wood discoloration	(a) Avoid wounding tree barks (b) Farm sanitation
Sap-stain (*Ilaga macana*)	*Ceratocystis fimbriata*	Foliage and trunk	Minor	Yellow leaves, wilting and dry leaves, dying branches, darkening of the wood	(a) Adequate pruning (b) Farm sanitation (c) Maintain soil fertility
Root-rots (a) *Rosellinia* root (b) *Armillaria* root (bark splitting)	(a) *Rosellinia bunodes*; *R. arcuata*; *R. pepo*; *R. necatrix*	Roots	Minor	Common on coffee planted on virgin forest soils. Precipitate weekening and wilting of trees, internal discoloration of roots to death of tree	(a) Farm sanitation (b) Trenching (c) Removal of dead wood from the farm
	(b) *Armillaria mellea*	Roots	Minor		
Root and wood rot	*Fomes applanatus* *F. lamoensis* *F. lignosus* *Ganderma pseudoferreum* *Rhizoctonia* sp.	Roots	Minor	Root decay	(a) Farm sanitation (b) Use fungicides
Damping-off	(a) *Pellicularia filamentosa* (b) *Rhizoctonia solani*	Seedling disease in the nursery	Minor but could be serious	Damping-off of seedling	(a) Reduce watering or drain seed beds (b) Use seed dressing
Miscellaneous Nursery diseases	*Hemileia vastatrix* *Cercospora coffeicola* *Glomerella cingulata*	Seedbed diseases	Minor	Wilting and dying of seedlings	(a) Farm sanitation

9.3 COFFEE DISEASES

With judicious pruning and good farm hygiene (adequate weeding, de-suckering, removal of dead branches and twigs) incidence of diseases on coffee is usually low.

The major diseases of coffee in West Africa are listed in Table 28.

9.4 DEFICIENCY SYMPTOMS IN COFFEE

Coffee suffers from the deficiency of a number of minerals. The main deficiency diseases in coffee are caused by shortages of iron, zinc, manganese and magnesium.

1. *Iron deficiency* Common with coffee, but not in West Africa. The most noticeable symptom is leaf chlorosis with green veins.

2. *Zinc deficiency* Symptoms include weakening of trees, chlorotic leaves with green veins, shortened internodes and bunching of leaves. This deficiency can be controlled by spraying the plants with foliar zinc fertilizers.

3. *Manganese deficiency* Symptoms of manganese deficiency in coffee are often seen. The leaf lamina displays intense yellowing with whitish specking, the midrib and veins remaining green. This appears as a reticulated green network over a yellow background. Correct by spraying affected trees with manganese sulphate and hydrated lime (2:1).

4. *Magnesium deficiency* Most common in heavily bearing coffee trees and most easily observed during harvest time. The characteristic symptoms are dull leaves, chlorosis between veins, and very light yellowing which may develop to ochre in severe cases. The midrib is usually yellow often with a narrow green zone. Affected trees usually defoliate after harvest.

Some degree of control can be achieved by spraying affected trees with dilute solution of magnesium sulphate. The sulphate can be given as an injection.

In addition to these major deficiencies, coffee reacts unfavourably to deficiencies of copper, phosphorus, sulphur and calcium.

9.5 PESTS OF COFFEE

Coffee berries on ripening are attractively coloured, numerous, and not concealed. This attracts insects; however, the colour and aroma of the ripening berries also attract birds. In West Africa some birds feed on coffee berries, the birds chipping off the pulp while the seeds are discarded. Also, rats and rabbits may feed on ripe coffee berries. Usually, they feed on the pulp while the seeds are discarded.

More serious are monkeys, which constitute a very big menace in coffee

Table 29 Insect pests of coffee in West Africa

Common name	Scientific name	Family	Pest status	Damage done	Control measures
Defoliator	*Epicampoptera glauca* Hmps	Drepanidae	Major	Larvae or catterpillars defoliate the tree	The larvae can be controlled by using Gammalin BHC
Berry borers	*Stenphanoderes hampei* Ferr	Scolytidae	Major	Beetles bore small holes in the ends of berries. Larvae feed and develop inside berry destroying it	Regular harvesting, and hygienic measures necessary. Spray with Gammalin 20 at least 2 applications during planting
Shot-hole borer	*Xyleborus compactus* Eichh	Scolytidae	Major	Beetles bore into fruits and branches of seedlings and trees of all ages. The bored holes are pin head in size and may be exuding sap or have a white streak of dried sap below the hole. The injury causes twigs to die back to the point of attack	Difficult to combat. Sevin 50 being a systemic insecticide may reduce population
Mole-crickets	*Gryllotalpa africana* P.d.B	Gryllotalpidae	Minor	Mole-crickets eat seedling plants at or below soil level causing loss of stand	Cover the plant with wire netting. Hand pick the insect
Wood eaters	*Nasutitermes* sp.	Termitidae	Major/minor	Termites tunnel into roots and trunks of trees and seedling plants causing destruction of the plant	Spraying Aldrin or Dieldrin (0.25% to 0.06%)
Variegated locust	*Zonocerus variegatus*	Acrididae	Minor	General feeder insect pierces and sucks on leaves, stems and fruits	Spray with Gammalin 20
Mealybug	*Planococcus kenyae*	Pseudococcidae		Secretes honey dew from which moulds grow	

cultivation. Monkeys harvest ripe berries in large quantities and feed on both the pulp and the seeds. They also cause much breakage when moving through the coffee bushes.

Insects, especially those that attack coffee berries, constitute the greatest danger to coffee cultivation. In West Africa, the most important pest of coffee is the 'coffee berry borer'. *Stephanoderes hampei* is African in origin, but it now occurs in all coffee growing countries of the world. This insect attacks coffee berries, bores into them, eats the contents leaving the berry empty. Coffee berry borer is very difficult to control once it has entered the berry. Control has been attempted through the use of systemic insecticides.

Other pests that are of importance in relation to coffee cultivation in West Africa are tabulated in Table 29.

9.6 COFFEE HARVESTING AND PROCESSING

9.6.1 Harvesting

Coffee plants come into bearing two to three years after field planting. The earliness of bearing depends on the variety, the suitability of the environment and adequate management. The fruits of coffee are known as berries. The immature fruits are green while the mature fruits are yellow, varying in some varieties from yellow to purple. The berries occur in clusters. Harvesting must be carried out when the berries are ripe. In all coffee producing countries, coffee harvesting is carried out by hand. It is an expensive operation. Where available, coffee harvesting is better carried out with children and women labour. Enforcement of tasks in coffee harvesting is very risky. The usual practice is to pay harvesters a minimum basic wage and adopt a system of bonus payment for hard work and efficiency of harvesting. Each batch of harvested produce brought to the collection centre by each harvester should be examined for unripe berries, wood and leaf fragments and other foreign matter. All contaminants should be removed before bulking the harvested produce prior to processing.

Coffee processing may be conveniently divided into two phases, a pre-industrial and an industrial phase.

9.6.2 Processing

9.6.2.1 Pre-industrial processing

The quality of coffee beans is very sensitive for the processing method used to prepare the crop. The two processes currently in use are the wet and dry methods.

9.6.2.1a Wet method

In this case a processing plant is required. To obtain good quality coffee, berries must be picked as soon as they are ripe and they must be de-pulped on the day of picking, as any delay causes heating through fermentation and this spoils the flavour of the bean.

The berries on their arrival at the processing plant are placed in cement tanks full of water where the first separation by density eliminates infested berries and foreign matter. The good and fully matured berries are then fed through a water channel into a pulping machine which removes the outer layer of the fruit. From the pulper the beans, coated with a mucilaginous covering, are carried down another channel to the fermentation tanks where they are left to ferment for periods varying with the ambient temperature. An improved method involves washing the beans with water under pressure to remove the mucilage and render fermentation unnecessary. When fermented or treated as described above, beans in the wet parchment skin are thoroughly washed and spread out on barbecues or dried artificially in ovens. Correct fermentation is the most important factor in the whole process of preparing the beans for the market.

9.6.2.1b Dry method

In this method, the berries are spread out in the sun on barbecues either at a central station used by several growers or each grower has his own drying surface which can be made of wood, earth-block or cement, on which the berries are exposed until dried. This process requires a climate with optimal drying conditions for a period of 10 to 15 days for each batch. After drying, the beans are hulled to remove the pericarp. Hulling can be done with the aid of hulling machines although it is usually done with pestle and mortar. Pericarp chaff is removed by winnowing, followed by picking. The efficiency of winnowing appears to be correlated with the efficiency of drying as this determines the breaking and falling apart of the pericarp from the coffee parchment. On the whole, parchment produced by the dry method is generally inferior to that produced by the wet method.

Advantages of the dry method are:

1. Its simplicity and relatively low capital cost per unit;
2. No need for people with special training for operating a plant.

Disadvantages are:

1. The long period of time needed for drying and the dependence on weather conditions;
2. The lower quality when compared to material produced by the 'wet' method;
3. Wide fluctuations in quality between growers.

While coffee processed by the wet method is, on the whole, considered to be of better cup quality than that processed by the dry method, yet some of the coffee produced in Yemen and certain areas in Brazil such as Sao Paulo, Ribeirao and Preto, are of excellent quality.

As a rule in West Africa arabica coffee is processed by the wet method while robusta coffee is processed by the dry method.

9.6.2.2 Industrial processing

The operations involved in the industrial phase of processing are grading, roasting and grinding.

9.6.2.2a Grading

Grading consists of about eight distinct operations, namely:

1. Re-drying – to ensure a uniform moisture content.
2. Cleaning – removal of any unhulled beans and foreign matter.
3. Hulling and polishing – primarily removal of the silverskin (testa).
4. Size grading – separation of broken beans and those of different sizes with a set of cylindrical sieves.
5. Density separation – separation of light black and defective beans from good coffee.
6. Hand sorting – removal of defective beans which the different machines could not eliminate.
7. Mixing – proportional mixing of clean size-graded coffee.
8. Bagging – sewing up mixed lots of specific weights for each roasting operation.

9.6.2.2b Roasting

Before the full flavour of the coffee can be brought out, the beans must be roasted for 20 to 30 minutes, though the time naturally varies with the variety and type of coffee and the degree of roasting required in the final product.

All machines, whether large or small, depend on the same principle: that of a rotating, closed cylinder which is heated. The cylinder must be closed because during roasting, volatile material is driven off and re-absorption of this by the beans later, gives the characteristic flavour. Small-scale hand-operated roasting drums taking a few kilos of beans at a time, and heated by a charcoal fire underneath, were produced in Nigeria during the Second World War and were found very efficient in producing a drinkable product from robusta beans.

9.6.2.2c Grinding

The roasted coffee has to be ground into small particles before it can be used. This is done by machines which again vary in size but which employ the principle of corrugated rollers with one rotating and the other stationary. Ground coffee especially when finely ground loses its flavour rapidly; it is therefore often sealed hermetically in tins before dispatch.

'Soluble' coffees, like Nescafe, are made from an infusion of coffee from roasted and ground beans, which is then dried very rapidly by spraying into very hot air. Extremely rapid evaporation takes place and the dried product easily dissolves again in water to give back the original infusion.

9.7 COFFEE BREEDING

Work on coffee breeding has had a short existence in West Africa. Most of the work has been carried out in Nigeria and the Ivory Coast. Some aspects of the coffee breeding programme are described.

9.7.1 Early Yield Trials

Early research work on coffee varieties introduced into West Africa involved variety trials at Ibadan, Umuahia and Owena in Nigeria, and Bingerville in the Ivory Coast. The most comprehensive trial lasted for almost 25 years and compared the performance of the Gold Coast, Java, Uganda and Quillou robustas (*Coffea canephora*). It was found that except in the first few years, Quillou robusta gave by far the highest yields of all. During the first 20 years, the order was Quillou, Java, Gold Coast and Uganda. After then, while Quillou remained the best yielder, Gold Coast robusta displaced Java for the second best. The superiority of Quillou over all other robustas in terms of yield has been confirmed at different locations. As a result of these trials, the highest yielding CRIN selections of robusta are Quillou clones C36, C96, C105 and C111, and Java clones T45, T129, T171, T197, T220, T395 and T1049.

Diallele crosses between these selections have been made to develop improved material. However, while yield is one of the first selection criteria, other factors considered and scored include regularity of bearing, habit of trees, size of bean, ratio of clean bean to fresh berries.

9.7.2 Germplasm Collections

Very little work has been done on coffee breeding in West Africa, although there is much potential. Germplasm of both the robusta- and arabica-type coffees is available in West Africa (Mambilla Plateau, Nigeria; Bingerville, Ivory Coast).

Collections of *Coffea canephora* and other diploid coffee species include Quillou, Java, Gold Coast, Plantation Java, Uganda and Niaouli varieties of robusta, and *C. stenophylla, C. excelsa* and *C. liberica*. These are all available in Nigerian collections. The robusta varieties in these collections are employed in progeny, variety and clonal trials.

Arrangements are under way for the introduction of a strain of caffeineless coffee and a high yielding robusta, selection S795 from the South India Coffee Research Station. Recent introductions from that station included arabica-robusta hybrids (triploids). Arabica-robusta hybrids have also been produced in Ivory Coast.

Farmers' cultivation of arabica coffee has been relatively recent. Most of the planting is done on the Mambilla area in north-east Nigeria with seeds from locally cultivated varieties introduced from Cameroun. In 1964, three high yielding varieties were introduced from Kenya: SL-14, SL-19 and SL-34. They were planted with a local variety at Gembu. Results of the first harvests showed that all three varieties yielded significantly higher than the local variety. The five best entries (including the local variety) were entered into a diallele crossing programme to develop improved material. These are being scored for rust resistance and other factors.

9.7.3 Kususku Arabica Gene Pool

Seeds of 115 varieties of rust resistant arabica coffee were introduced in 1965 for establishing a gene pool at Mambilla. Seedlings were raised and planted out in 1966.

Flowering first occurred in 1969. Records of vegetative growth, periodicity of leaf flushing, heights, number of laterals, scores on rust resistance have been taken. Yield records include conversion factors of berry to parchment. Comprehensive records covering all plants in all surviving varieties are taken and the varieties then assessed for their performances. Finally trees from the most outstanding varieties are included in a hybridization programme, for use as recurrent parents for combining rust resistance with high yield.

9.8 OUTLOOK ON THE FUTURE

The greatest problems to be encountered in coffee in West Africa include the following:

1. *Non-uniform ripening of berries* This is a problem of both the arabica and canephora-type coffees. It is a very serious defect in coffee because it increases the harvesting costs, harvesters having to visit a single stand several times before completing harvesting; the inclusion of immature berries in the produce also lowers the quality of produce.

2. *Development of appropriate pruning system* Coffee pruning is very little developed in West Africa. Inadequate pruning has lowered the yield of coffee in the producing countries. More effort is needed in coffee pruning.

3. *Use of the by-products of coffee* The coffee berry husk constitutes about 30 per cent by weight of the wet berry. So far the husk is a wasted product in Africa. The husk is a potentially valuable livestock feed if properly decaffeinated. A promising approach to coffee breeding that could yield handsome results is the production of interspecific triploid hybirds between *C. arabica* and *C. canephora*, thus combining in a single hybird the arabica qualities with the robusta vigour and resistance to diseases. The problem of sterility in such hybirds can be overcome through chromosome doubling.

ADDITIONAL READING

Stolezenbach, C. (1979). Reafforestation with Indigenous Tree Species of a Chilean State Forest. *Plant Research and Development*, **10**, 105–115.

Fredo, O. *et al.* (1980). Research into the Ecophysiology of Brazilian Tree Species as a Basis for the Management of Natural Forests. *Plant Research and Development*, **12**, 69–75.

Wellman, F. L. (1961). *World Crops Books Coffea; Botany, Cultivation and Utilization.* Leonard Hill Books Ltd. London.

Wrigley, G. (1971). *Tropical Agriculture: The Development of Production.* Faber and Faber, London.

Sauds, P. B. (1968). *Coffee Production in Western Nigeria.* Government Printers, Ibadan, Nigeria.

CHAPTER 10

Citrus Fruits

Citrus species probably originated in north-eastern India, in Burma and in the adjoining areas. Early in the spread of citrus, some species crossed into China where the sweet orange, the mandarins and the kumquat developed. The lemons, the limes and the grapefruit appear to have developed and spread from India to the Near East and to the Mediterranean region. Citrus species were carried to other parts of the world from the Mediterranean region.

Spain and Portugal were credited with the honour of distributing various species of citrus to their colonies in America and Africa for planting. There are, however, no accurate records to show when the various species of citrus arrived in West Africa.

Citrus species are grown principally for the juices of their fruits. After extraction of the juice, the fruit pulp is a possible livestock feed; the rind acid (oil) is an expensive commodity in the international market. Citrus seeds are known to contain sweetening agents which are being studied as a probable substitute for sugar in the world market.

Citrus cultivation is on a small scale in West Africa. Most citrus fruits produced are consumed locally. The cultivation of citrus is yet to be organized and encouraged so as to maximize the use of the various products obtainable from the fruits.

10.1 TAXONOMY

The commonly grown *Citrus* species belong to the family Rutaceae which contains about 150 genera and nearly 2000 species. The genus *Citrus* contains all the species widely cultivated in West Africa.

The most important *Citrus* species grown in West Africa are as follows:

Sweet orange	*Citrus sinensis*
Sour orange	*Citrus aurantium*
Lime	*Citrus aurantifolia*
Lemon	*Citrus limon*
King orange	*Citrus sinensis* × *Citrus reticulata*
Tangerine, Mandarin	*Citrus reticulata*
Grapefruit	*Citrus paradisi*
Shaddock (pomelo)	*Citrus grandis*
Tangelo	*Citrus paradisi* × *Citrus reticulata* (an interspecific hybrid)

Citrus species are evergreen trees of small to medium stature. They often have thorny (prickly) stems. The leaves are unifoliate, petiolated and quite often the petioles are winged. Flowers are perfect, usually large, fragrant, mostly white. The leaves, twigs and rind of the fruits contain oil glands which secrete the rind oil (acid) of commerce (see Figure 33).

Citrus fruits are small to large with leathery rind, yellow to orange in colour when ripe, pulp and juice may vary in taste from sweet to acid, the fruit segments may vary from 8 to 18, but usually are 10 to 14. Seeds may be many, few, or none as in the tangelo.

Flowers are carried singly or in small clusters; they are sweet scented, and with five-membered perianth segments. Stamens are numerous, usually assembled into five groups. The filaments of each group are partially united at the base. The ovary is superior, with axile placentation. It is surmounted by a short style with a globular stigma which bears as many passages as there are loculi. The style is deciduous. Generally in citrus the androecium and the gynaecium mature at the same time. However, some degree of protandry has been exhibited in some newly bred varieties of citrus. Pollination in citrus is brought about by insects especially the bees. The pollinating insects are attracted to the flowers by the conspicuous corolla, the abundant nectar and the strong scent produced by the trees. Pollen grains are sticky and easily transplanted by insects. Cross-pollination is generally the case, but selfing occasionally occurs.

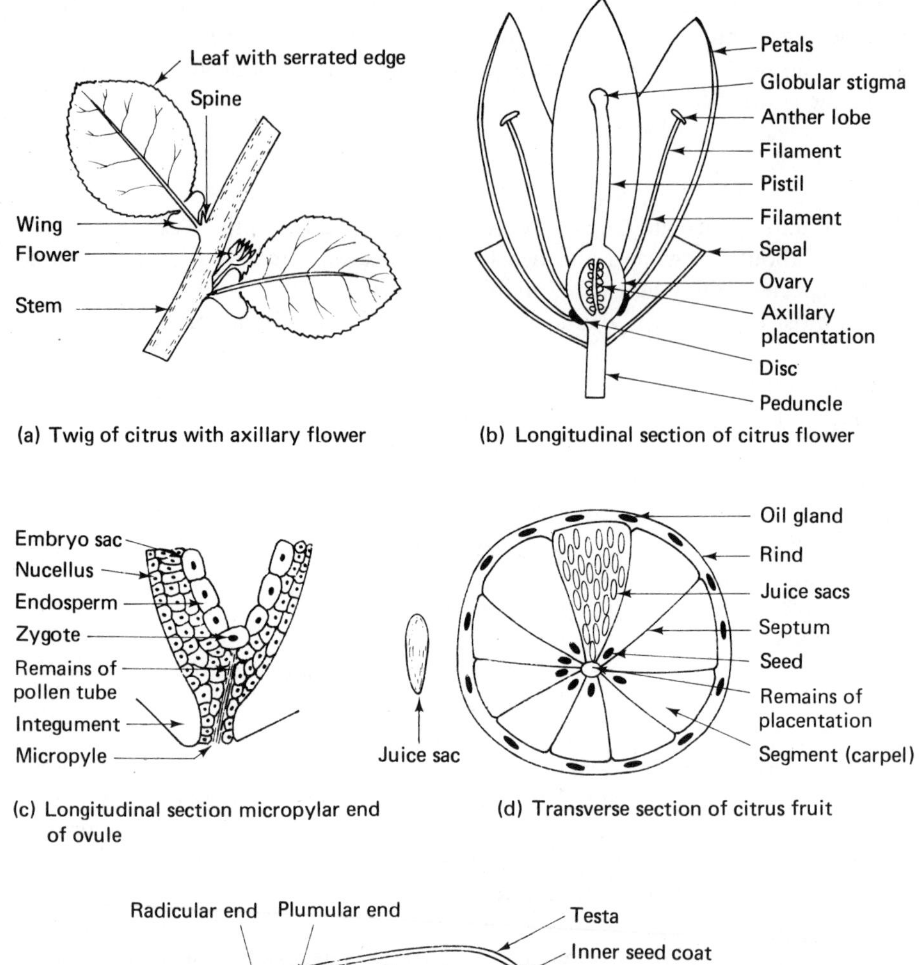

(a) Twig of citrus with axillary flower

(b) Longitudinal section of citrus flower

(c) Longitudinal section micropylar end of ovule

(d) Transverse section of citrus fruit

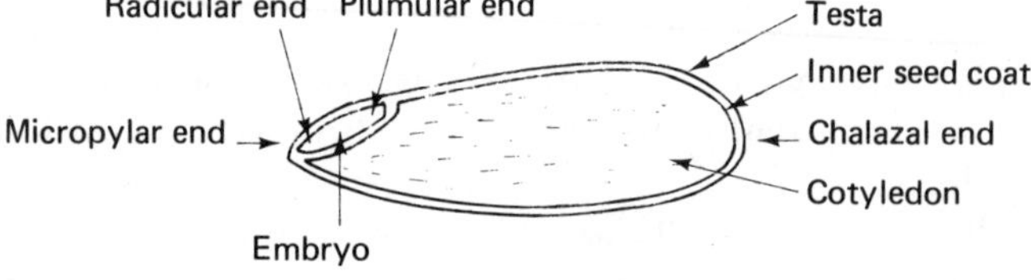

(e) Longitudinal section of citrus seed

Figure 33. Morphological features of citrus

A number of peculiar biological phenomena occur in the production of citrus. The first of these is the occurrence of nucellus (nucellar) embryos. Nucellar embryos form a type of apomixis in which one or more embryos arise from a group of cells either in the nucellus or from the integuments of the ovule as a result of chemical stimulation by the process of fertilization, which occurs in the adjacent gametes. Nucellar embryos are diploid and contain the maternal set of chromosomes. When, however, such embryos arise from the synergids, the apomictic embryo is haploid. Such haploid embryos have been used to produce diploid isogenic lines of citrus. In citrus, therefore, normal fertilization takes place to produce the sexual embryos, but often nucellar embryos which are maternally diploid arise in the nucellus or integument tissues of the ovule. As a result, there is frequent occurrence of polyembryony.

The fruit of the citrus is a berry in which the exocarp and the mesocarp are leathery in texture. It is known as a hesperidium. The epidermis of the fruit is covered with a thick cuticle. Stomata occur in the epidermal layers of the cells. The exocarp is a thick layer of irregularly shaped parenchyma cells, it is rich in chloroplasts and contributes appreciably to the photosynthetic activity of the leaves. There are numerous oil glands in the exocarpic tissue. The ductless glands are filled with essential oil known as citrus rind oil in the trade. The spectrum of the oil varies from one species of citrus to the other, but the extraction of the oil is an essential side enterprise of the citrus industry.

Citrus seeds occur within the berry and are embedded in the juice sacs of each loculus, very close to the central axis. Chemically, the citrus juice on average contains, on dry matter basis, about 90 per cent sugars and acids. As the fruit ripens, the chlorophyl contained in the chloroplastids of the exocarp breaks down and the volume of juice within the fruit increases; there is also a small reduction in the acid content of the juice.

The family Rutaceae is renowned for the production of essential oils. The citrus group produces different types of essential oil in the fruit rind. Some of these oils are lime oil, sweet orange oil, bergamot oil obtainable from *Citrus aurantium*, lemon oil and neroli oil (obtained by the distillation of the flowers of bitter and sweet orange).

10.2 CITRUS AND ITS ENVIRONMENT

All cultivated citrus species grow very well in both the tropical and subtropical parts of the world provided there is sufficient moisture and the temperature does not drop below freezing point. The main produce from citrus is water-based and, therefore, adequate supply of moisture especially during fruit development is very important. Citrus fruits tend to be small (undersized) when grown under dry or semi-dry conditions. Alternatively, excessive moisture is not conducive to optimum productivity of citrus. When citrus fruits develop under excessive moisture, the juice tends to be 'watery' and the fruits oversized and puffy. Citrus grows best under 1100 to 1500 mm of rain per annum, distributed over nine months of the year.

The temperature range for good growth and productivity varies from 13°C to 37 °C with the optimum at 28 °C. The optimum temperature for maturation of best quality fruits is 13 °C to 17 °C. This special temperature requirement for maturation explains why citrus fruits which mature under subtropical and Mediterranean conditions are superior in quality to those that mature under high tropical temperature. Citrus requires a period of dry but cool weather for final ripening of the fruits. Excessive sky cover is not favourable for good growth and production of citrus.

10.3 AGRONOMY

The principles of cultivating citrus follow the same pattern as for other tree crops in West Africa. In the discussion that follows, operations that are the same are mentioned only briefly.

10.3.1 Selection of a Site

All citrus species will grow on a wide range of soils. As the taproot plays an important role, the soil must be deep, free from iron concretions, well drained and high in fertility. High organic matter is known to favour good growth and production. Any site selected should be level or only slightly sloping. It should also be protected from strong winds either naturally (aspect!) or by the establishment of windbreaks.

10.3.2 Land Preparation

Land preparation for planting citrus consists of clear-felling of any vegetation. Any excess trash should be burnt. If felling can be arranged a year ahead of establishing the citrus orchard a leguminous fallow can be established already at that time. Otherwise this is to be sown at the time of planting the orchard. The area is to be lined out with planting sites at 6×6 m or 7×7 m. The usual size of the planting holes is $60 \times 60 \times 60$ cm.

10.3.3 Cropping Pattern

Citrus orchards are usually established as monocrops, rarely mixed with other citrus varieties or even other citrus species. Nevertheless, a mixture of different citrus species is sometimes practised.

Sometimes, citrus is intercropped with other types of crops such as pawpaw, pineapple, vegetables, cereals and tubers during the first two to four years after planting in the field and before the citrus plants develop a full canopy. Intercropping should be stopped when the citrus develops its canopy. Application of fertilizers is recommended when an intercropping programme is adopted. It is advisable to avoid bananas and/or plantains as intercrops in citrus orchards, as they are difficult to eradicate when no longer needed.

10.3.4 Plant Propagation

10.3.4.1 Propagation by seed

10.3.4.1a Seed collection

Plants which are to serve as a source of seeds for planting should be selected on the basis of high yield, good growth, freedom from pests and diseases, especially virus infections, and should give good quality produce, e.g. juicy fruits with good flavour. When fruits are harvested for the purpose of extracting the seeds for planting, the fruits should be selected on the basis of size, fullness and conformation. Seeds also should be selected on the basis of size – the larger the seed, the larger generally the embryo and the food store and the more vigorous the germination and early establishment.

Seeds should only be taken from ripe and well-filled fruits. Freshly extracted seeds should be sown immediately, as citrus seeds rapidly loose viability on extraction, and during storage under normal conditions. If it becomes necessary to store citrus seeds for a short period of time, the seeds should be kept in the fruit. If this is not possible the following procedure should be followed:

1. Extract seeds.
2. Mix the freshly extracted seeds with moist fine sand, sawdust or moist ground charcoal.
3. Store in a closed jar in a *cool* place.

Under these conditions, extracted citrus seeds can be stored for 10 to 15 days without loss of viability.

10.3.4.1b Sowing

Although citrus seeds can be sown at stake, it has now become an accepted practice to raise citrus seedlings in the nursery either for transplanting to the field as seedlings or for use as stocks in budding. Citrus seeds are sown either in seed boxes or in nursery beds as single seeds spaced at 3 × 6 cm. To allow good growth seedlings should be pricked when they are 3 to 6 cm in height into either 15 × 24 cm cane baskets or 15 × 24 cm polybags which have been filled with fertile topsoil. Seedlings can also be pricked to nursery beds at a spacing of 15 × 30 cm. Seedlings should be selected for vigour of growth; weak seedlings should be discarded. Although citrus seedlings do not require shade either in the nursery or in the field it is a general practice to provide sown seeds with light shade until the time of pricking. After pricking the seedlings are grown without shade. It is essential to ensure adequate watering, mulching, pruning (removal of laterals), pest and disease control and adequate nutrition of plants from germination throughout the life of the plants.

10.3.4.2 Vegetative propagation

The most common method of vegetative propagation used with citrus is budding. The inverted 'T' and the normal 'T' methods of budding are used. The seedlings to be used as root stocks are raised in the same manner to the age of 12 to 15 months after pricking or to a height of at least 45 cm.

When the root stock seedlings have attained this size, or when budding is due, they are prepared for the budding operation. This consists of pruning (including the removal of leaves) the seedlings to a height of 45 cm a week before the budding operation. Budwood is normally collected on the morning of the day of budding and used immediately. In case budwood has to be stored for transportation, it should be packed in moist cotton-wool, wrapped in moist sacking, firmly tied and stored in a cool place. In this way, budwood can remain fresh for up to seven days. Budwood is usually taken from branches which are 8 to 10 months old. Budding is carried out in the cool hours of the morning or in the evening at a height of 25 to 30 cm from ground level on the prepared seedling root stock. Budding should be avoided during the hot hours of the day and during heavy rains.

The budded stocks are looked after in the normal way until they are old enough for transplanting into the field.

10.3.5 Root Stocks and Their Use in Citri-culture

Root stocks are seedlings raised for the purpose of supporting the scion budded or grafted on to the root stock. In general, some root stocks give quality improvement effects to the scions. Njoku and Obasi (1976) studied the influence of citrus stocks on yield, fruit quality and nutrient uptake of Nigerian sweet orange, at Ibadan, Nigeria, using: (*a*) Lake tangelo; (*b*) Cleopatra mandarin; (*c*) Sampson tangelo; (*d*) Sierra Leone rough lemon; (*e*) Nigerian sweet orange; (*f*) Nigerian sour orange stocks for Nigerian sweet orange scions. The control was Nigerian sweet orange seedling. These authors found that out of the six root stocks tested, Lake tangelo followed by Cleopatra mandarin by far outyielded Sampson tangelo, Sierra Leone rough lemon, Nigerian sweet orange, and Nigerian sweet orange seedlings. The juice content was not affected by the type of root stock but fruits from Lake tangelo and Sierra Leone rough lemon had higher acid contents while Cleopatra mandarin and Nigerian sweet orange stocks produced fruits with the highest total soluble solids. The Nigerian sweet orange, especially the seedlings, tended to produce small fruits while fruits from Sierra Leone rough lemon and Lake tangelo root stocks were relatively large. It was also found that the type of root stock markedly affected the uptake of P, K, Ca, Mn and Cu, but that the Nigerian sweet orange seedling exhibited the lowest ability to take up nutrients from the soil.

Special root stocks have been used in citrus culture in Nigeria as a means of disease control, especially the control of Tristeza, a deadly virus disease of citrus. Root stocks that have been shown to be resistant to Tristeza in Nigeria are, in order of importance: rough lemon, Cleopatra mandarin, willow-leaf-Tangerine, and Lake tangelo. The major criteria for the selection of root stocks are the ability to withstand or resist the more prevalent citrus disease in the locality. In most parts of West Africa, the two most dreaded diseases are tristeza and gummosis. Furthermore, good root stocks should maintain good growth and vigour, support high yields of good quality and maintain even flowering and maturity. That root stocks may influence scion characteristics is shown in

Table 30 Mean nutrient contents of sweet orange leaves as affected by various root stocks

	N	P	K	Ca	Mg	Mn	Cu	Zn	Fe
Root stocks	Per cent					p.p.m.			
1. Sampson tangelo	2.22	0.113	1.50	2.75	0.39	28	8.6	6.0	158
2. Lake tangelo	2.41	0.117	1.10	3.65	0.50	20	7.0	6.0	168
3. Cleopatra mandarin	2.33	0.141	0.73	3.54	0.55	59	5.0	6.0	135
4. Sierra Leone rough lemon	2.33	0.131	0.86	3.26	0.53	59	5.0	7.0	195
5. Nigerian sweet orange:									
(seedling)	2.15	0.136	0.71	2.65	0.37	29	7.0	4.0	170
(adult)	2.24	0.130	0.97	1.88	0.49	43	9.8	6.3	138
LSD	0.165	0.023	0.23	0.853	0.340	15.6	1.40	4.16	34.06
SE	0.0564	0.0079	0.0792	0.2914	0.1163	5.33	0.479	1.424	11.640

Source: B. O. Njoku and N. K. Obasi, 1976. *Nigerian Agricultural Journal*, **13**, 54–64.

Table 30, indicating differing nutrient levels in leaves of a sweet orange scion on four root stocks or on its own roots.

10.3.6 Transplanting

The transplanting of citrus follows the standard procedure for transplanting tree crop seedlings (see Chapter 7).

10.3.7 Post-planting Maintenance Operations

The essential post-planting maintenance operations that should be carried out in a citrus orchard are as follows:

1. Establishment of a ground cover of legumes. The ground cover could be established before citrus seedlings are transplanted into the field or immediately after transplanting.
2. Mulching of each seedling towards the end of the rains.
3. Removal of undecomposed mulch during the early rains.
4. Maintenance of the leguminous ground cover – weeding and re-seeding of gaps.
5. Watering/irrigation as necessary.
6. Pruning to shape the trees and removal of suckers and laterals. Removal of laterals must be regularly carried out especially in orchards planted with budded material.
7. Application of fertilizer. Fertilizer application should be carried out in May and September, i.e. during the long and the short rains. Only fertilizer mixtures recommended by the soil scientists should be used.
8. Regular control of pests and diseases through farm sanitation, use of resistant varieties and application of insecticides and fungicides.
9. Regular weed control either by hand weeding or use of herbicides.

When seedlings come into bearing three to four years after planting, there is a tendency for fruit-bearing branches to assume horizontal positions and later bend downwards. This may break the fruiting branches and to avoid such damage, heavily laden fruiting branches should be supported with forked sticks.

In areas where there is shortage of agricultural land and in cases where orchards are located near cities, it is normal practice to intercrop newly planted citrus orchards with vegetables. In this case a leguminous ground cover is not planted until intercropping is stopped. In such cases, intercropping should be stopped after 24 months of field planting.

10.3.8 Production and Harvesting

Citrus begins to bear from three to five years after planting in the field depending on the species, variety and method of propagation. Vegetatively propagated material comes into bearing earlier than seedlings.

The main commercial product of citrus is the fruit juice. The citrus industry of the world is principally aimed at the production of fruits for juice extraction in various forms for consumption. In addition to the juice, the following by-products are obtainable from citrus.

1. The rind oil is used in pharmaceutical industries.
2. Fruit pulp is the mesocarp tissue which is left as residue after the juice has been extracted. Where available in large quantities it is used as valuable livestock feed ingredient. It is also used as fuel.
3. The wood of citrus plants is an excellent source of firewood. It is also used in thorn carving.

In citrus, the colour of the fruit rind is green when fruits are not mature. On maturity, the fruits begin to ripen during which period the rind colour turns yellow.

The quality of citrus is determined primarily by the taste, juiciness and size of the fruit. In addition, the quality is affected by the trueness to type of the fruit. In the world market, however, the marketing value is determined to a large extent by the uniformity of the yellowness of the rind on ripening.

Citrus fruits are easily perishable after ripening. To ensure that fruits are not lost through over-ripening, fruits should be harvested immediately they are ripe and delivered to consumers, processors or retailers. Fruits should be harvested with optimum care to ensure that they are not bruised or damaged in any form. Harvesting operations should also not cause any damage to the trees. After harvesting, fruits should be graded into sizes with diseased/broken and rotten ones removed and destroyed. In some cases, fruits may need to be stored for some time while awaiting markets or transportation to distant markets. Under such circumstances, fruits are harvested a short time before full ripening, the time interval between harvesting and full ripening depends on the length of time the fruits are to be stored or the length of time required for transportation. It is, however, essential that fruits are harvested only when they are physiologically mature.

It is a common sight in West Africa to find citrus fruits loaded into lorries whose slatted wooded bodies have been lined with oil palm or coconut leaves. Fruits transported in this way suffer considerable damage through bruising. The correct method for packing citrus fruits for transportation is to pack them in wooden crates of dimension 75 × 75 × 60 cm. For transportation over short distances, the tops of such crates need not be closed, but for long distance transportation, the crates should be closed. Storage and transportation of citrus fruits should always be carried out under cool conditions.

When fruits are destined for distant markets, they are harvested when physiologically mature but before ripening. Such fruits generally ripen en route to the market or they are treated with chemicals (ethylene, etc.) to ensure uniform ripening.

Most citrus fruits produced in West Africa are consumed fresh. Production is grossly below demand. Sahelian countries of West Africa depend on importation of citrus fruits from the Mediterranean countries.

A few citrus processing plants have been established in West Africa (Nigeria, Ghana and Ivory Coast). These factories are not in full production because of insufficient and irregular supply of fruits. There is an urgent need for West Africa to develop her citrus industry.

10.4 DISEASES OF CITRUS

Citrus diseases may be caused by virus, bacteria, fungi, algae, nematodes and plant parasites. Citrus plants are also known to suffer from a number of deficiency diseases. A few of the important diseases of citrus in West Africa are listed below:

10.4.1 Major Diseases

1. *Tristeza* A virus disease of citrus, usually very severe on sweet oranges, grapefruits, and tangerine budded into sour orange. It is sap transmissible. The most common method of control or prevention of the outbreak of Tristeza in West Africa is to use any of the following as root stock:

(a) Rough lemon
(b) Cleopatra mandarin
(c) Willow-leaf tangerine
(d) Lake tangelo.

2. *Anthracnose* Anthracnose in citrus is caused by species of *Gloeosporium* and *Colletotrichum*. The major symptoms are leaf blight, twig blight and fruit staining. Anthracnose could be very serious especially on varieties of lime. The most common methods of control are:

(a) Spraying with copper fungicides
(b) Farm sanitation
(c) Use of tolerant/resistant varieties.

3 *Scab* Scab in citrus is caused by species of *Elsinoe* and *Sphaceloma*. The main symptoms are whitish scabs on leaves, twigs and fruits of citrus. Sour orange is especially susceptible to scab. The major methods of control are:

(a) Farm sanitation
(b) Use of tolerant/resistant varieties
(c) Use of fungicides, e.g. Captan, Bordeaux mixture.

4. *Melanose* Melanose where it occurs is a very serious fungal disease of citrus. It is caused by species of *Phomopsis*. It kills affected leaves and twigs, while it stains and causes fruits to decay. Control is by spraying with fungicides. Resistant varieties are yet to be bred in West Africa.

5. *Foot rot or brown rot* Simply known as citrus gummosis. It is caused by *Phytophthora* sp. It kills the bark on trunks and roots eventually resulting in the death of citrus plants. In mild cases it results in fruit rot. Control is to treat plants – trunk, roots and twigs – with effective fungicides.

10.4.2 Minor Diseases

Citrus suffers from a number of minor diseases. In most cases, these minor diseases do not merit special control. Adequate farm sanitation has been shown to be effective in keeping them in check.

1. *Sooty mould* This is a blackish sooty fungus which grows on exudates from wounds caused on citrus trunks and twigs by insects. Sooty mould is common where citrus is shaded and where the citrus canopy is too dense.

2. *Algal leaf spots* These are algal species growing on the trunks, twigs, leaves and fruits. Ordinarily they cause very little damage. Control is through adequate farm sanitation.

3. *Mistletoe* This is an epiphytic parasite. It is a true green plant which grows on citrus twigs depriving the citrus of nutrients which it needs for its own photosynthetic activities.

4. *Mineral deficiencies* Citrus suffers from the deficiencies of nitrogen, phosphorus and potassium. The symptoms are the standard ones. In addition, citrus suffers from a shortage of zinc. The main symptom of zinc deficiency is sickle leaf. Iron deficiency is also common in citrus. The main symptom is chlorotic patches. All deficiency diseases can be easily treated by the application of the appropriate nutrient.

10.5 INSECT PESTS

1. *Purple scale* This occurs in large numbers on leaves and branches. It is easily controlled by spraying with an oil emulsion. It can also be controlled by use of its parasites (biological control).

2. *Mealybugs* Mealybugs usually cause injuries to the growing points of citrus. They are easily controlled by spraying with Parathion or Malathion.

3. *Aphids* Aphids feed on succulent growing shoots and leaves of citrus especially at the nursery stage. Aphids can be a nuisance on young citrus plants. The main control is to spray with nicotine sulphate or similar insecticides.

4. *Mites* Mites are common on citrus leaves during the dry season. They usually occupy the lower surfaces of leaves, causing russeting of leaves and fruits. Russeting can be controlled by spraying with Rogor 40.

5. *Fruit moths* Outbreak of fruit moths on citrus could be very serious. The adult moths puncture the fruits and succulent leaves; This is followed by invasion of saprophytic fungi which cause the fruits to rot and drop. Moths are difficult to control. When an outbreak of moth attack is suspected, an immediate report should be made to an entomologist for advice.

In spraying citrus with fungicides and insecticides, caution should be entertained. Certain persistent insecticides such as DDT could do more harm than good as they kill both pests, predators and parasites. It is a golden rule to consult a specialist before the application of chemical sprays to citrus.

ADDITIONAL READING

Braverman, J. B. S. (1949). *Citrus Products: Chemical Composition and Technology.* Interscience Publishers Inc. New York, USA.

Food For Us All (1969). *The Yearbook of Agriculture* (1969). United States Government Printing Office, USA.

Gardner Bradford and Hooker, J. 3rd Edition (1952). *The Fundamentals of Fruit Production*: McGraw Hill Book Co. Inc. London.

Childers, N. F. (1961). *Modern Fruit Science.* Horticultural Publications: Rutgers: the State University, New Jersey, USA.

Childers, N. F. (1966). *Temperate to Tropical Fruit Nutrition.* 174–228. Horticultural Publications: Rutgers, the State University, New Jersey, USA.

Njoku, B. O. and Obasi, N. K. (1976). *Nigerial Agricultural Journal*, **13**, 54–64.

CHAPTER 11

Latex Producing Plants

11.1 THE VARIOUS RUBBER PLANTS

The production of rubber as a crop followed the discovery of *Hevea braziliensis* as a crop of the Red Indians, the aborigines of Latin America, by Christopher Columbus and other Spanish explorers. Astronomer de la Condamine was the first to send samples of this mysterious elastic substance or 'Caout-chouc' to France from Peru in 1736. De la Condamine reported accurately on the trees which produce rubber. He also described methods of collecting the latex, of processing procedures and indicated its possible use.

On arrival in France, great publicity was given to the mysterious substance and a market was created in Europe. As a result of this, many plant species all over the world were examined and exploited for their rubber (resin, isoprene) content. The most important tree species that have been exploited for their rubber contents are:

Ficus elastica
Castilloa elastica
Funtumia elastica

Willughbeia sp.
Landolphia sp.
Palaquim gutta
Payena sp.
Mimusops balata
Achras zapota

Manihot glaziovii
Cryptostegia sp.
Parthenium argentatum (Guayule)
Solidago sp.
Hevea collina
Hevea guianensis
Hevea spruceana
Hevea braziliensis

The rubber industry started to develop fast from 1839 when Goodyear invented the vulcanization process, using sulphur. This discovery revolutionized the rubber industry. The invention resulted in the production of high quality rubber goods which lacked the deleterious effects of the raw rubber. Among the numerous inventions which followed the vulcanization process, the invention of the pneumatic tyre in 1888 gave the highest impetus to the rubber trade. From that time onwards demand for rubber rose very steeply and many latex-producing tree species were exploited.

Prior to 1900, the bulk of world rubber supply came from Brazil. The automobile was developed in 1895. This resulted in such a high demand for rubber that Brazil was unable to cope. Subsequently, many countries entered rubber production. The first export of rubber latex from West Africa was in 1880. The consignment was made up of 0.5 tonne (Ghana) and 450 tonnes (Sierra Leone). The first recorded export of latex rubber from Nigeria was in 1895, the consignment being 2322 tonnes.

The development of rubber cultivation in West Africa was due to the interest of European rubber traders. To encourage rubber planting, Firestone Tyre and Rubber Company negotiated for land for planting rubber in Liberia and other parts of West Africa as early as 1916. Seeds of *Hevea braziliensis* were introduced from Latin America. During the First World War (1914–1918), Britain was in dire need of rubber and she offered high prices for latex rubber. This supported further spread of the rubber plant in West Africa. To meet the high demand, *Funtumia elastica* (Lagos rubber) was heavily tapped all over West Africa. After the war, British trading companies started to establish rubber plantations and encouraged local farmers to plant rubber in different parts of West Africa, notably in the Bendel, Cross River and Rivers States of Nigeria; Ashanti in Ghana and south-western Sierra Leone. Local farmers took kindly to rubber cultivation. In addition to production by local farmers, a number of rubber plantations have been established in West Africa by the different governments, e.g. Ikenne, Araromi rubber estates in Nigeria. During the Second World War, a quantity of *F. elastica* was tapped. Since then the bulk of the world rubber has come from *Hevea braziliensis*.

11.2 HEVEA BRAZILIENSIS

11.2.1 Taxonomy and Botany

Hevea braziliensis is a member of the family Euphorbiaceae. The family also contains such plants as *Ricinus communis* (castor oil plant), *Manihot esculenta* (the cassava plant) and many other species of tropical importance. The genus *Hevea* contains a number of species but *H. braziliensis* is the only species with very high isoprene content in its latex. *Hevea braziliensis* is a native of the Amazon basin. It is a lowland tropical tree, thriving best under rainfall of 1900 to 2000 mm per annum and a temperature range of 24 °C to 32 °C.

H. braziliensis is an evergreen tree, 18 to 30 m in height when fully grown. It is straight with a smooth grey bark, canopy crown-like, and the branches are slender. Leaves alternate, palmate, and each leaf carries three leaflets. The leaflet is elliptic petiolated, with a basal gland, and pointed at the tip, with lengths varying up to 45 cm. Leaflets are glabrous, with entire margin and pinnate venation.

The inflorescences are in the form of pyramidal-shaped axillary panicles. They are produced simultaneously with new leaves, and they are arranged in a cymose form. Flowers are small, greenish white, dioecious; female flowers are usually larger than the male and more terminal on the branches, strongly scented, apetalous and with a five-lobed calyx tube. In the female flower, the gynaecium is composed of three united carpels, forming a three-lobed, three-celled ovary with a single ovule in each cell. The stigma is sessile and bifid. The male flower carries ten stamens arranged in two series of five, one set above the other on a central column. Pollination is carried out by moths, bees and flies and *H. braziliensis* is outbreeding. After pollination, the fruits mature in about six to

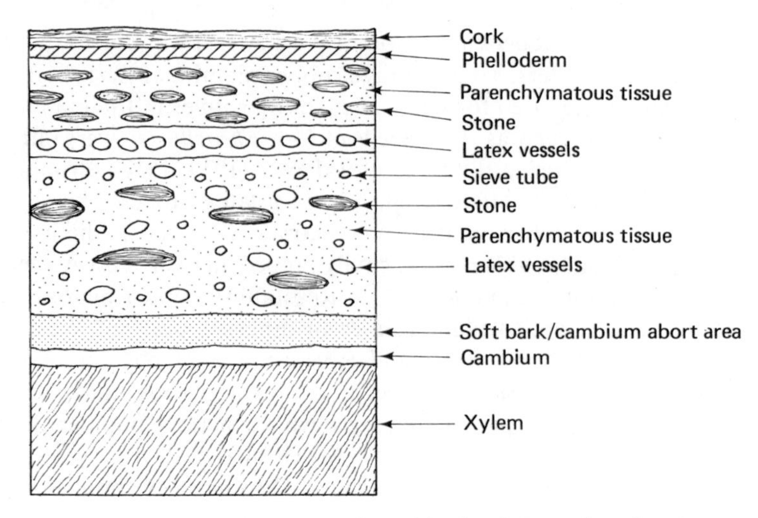

Figure 34. Transverse section of bark of *Hevea braziliensis*

seven months and the mature fruits dehisce explosively scattering the seeds some distance away from the mother trees. Seed is composed of a thick testa and soft kernel. The kernel contains a dark red drying oil, a lypolytic enzyme and a cyanogenetic glucoside. The oil, when extracted, is used for soap manufacture. The seeds lose viability very rapidly.

The bark of *H. braziliensis* is commercially the most important part of the tree (see Figure 34).

The mature bark is composed of:

1. *Cork* – the most outward layer which is brownish and smooth.
2. *Phelloderm* – thin and fragile.
3. *Hard bark* – this contains stones which decrease in density from outside inwards and between the stones are parenchymatous tissues.
4. *Latex vessels* – lie below the hard bark.
5. *Soft bark* – this is composed of parenchymatous tissues in layers alternating with smaller and more numerous groups of latex vessels. The soft bark latex vessels are less dense further out than towards the centre of the soft bark.
6. *Soft bark/cambium area* – where the soft bark and the outer cambial layers merge.
7. *Cambial layer* – layer responsible for the normal cambial activities of the plant.

11.2.1.1 The latex vessels

The latex vessels of *H. braziliensis* are excretory systems which are located in the cortex and are derived from the cambium. *Hevea* has a compound system

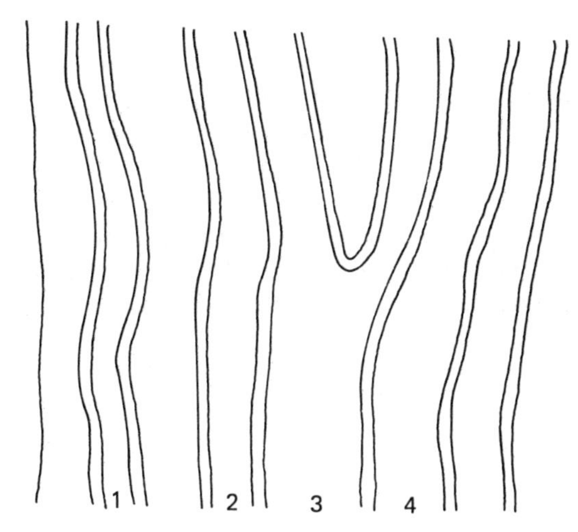

Figure 35. A cylinder of latex vessels

of latex vessels which originates through the fusion of adjacent cells whose cell walls disintegrate as the cells mature. When fusion of adjacent cells is complete, the nuclei of the original cells become grouped together at various points along the vessels. The main features of the latex vessels are as follows:

1. Cells which later become latex vessels are formed by the cambium in a controlled sequence alternating with the parenchyma cells. As a result, the latex vessels run in concentric circles or cylinders (see Figure 35).
2. Latex vessels in *H. braziliensis* are laterally (or transversely) interconnected. Latex vessels in the young cylinders run almost longitudinally.
3. As a tree of *H. braziliensis* expands in girth, the anastomoses of cylinders run almost tangentially. The tangential pull on the outermost latex vessel cylinders often causes disruption of the original anastomoses.
4. Radial connections between latex vessel cylinders occur; the number of such connections is usually small and varies from tree to tree.
5. In seedling plants, the number of latex vessel cylinders decreases in the younger parts of the stem.

11.2.1.2 Contribution of latex vessel cylinders to yield

In *Hevea braziliensis*, the contribution of the differently aged cylinders to latex yield has been intensively studied. All the studies that have been carried out show that the eldest and outermost latex vessel cylinders contribute very little to yield; the main latex production comes from the youngest latex vessel cylinders which are close to the cambial layer. As a result of girth growth, the outermost tissues of the bark undergo tangential pull. To withstand this pull, tangential cell elements arise from the surrounding paranchyma cells. The latex vessel elements in this area of the bark become pulled apart; new cells which do not assume a latex excreting function arise and these form stone cells.

The activity of the different layers of latex vessel cylinders has serious implications in successful tapping of rubber. The stone cells are originally parenchyma cells, which mostly originated from the cambium with a very small portion arising from the phelloderm. The ability to form stone cells is known to be hereditary and a study of stone cell formation has been used as a criterion for selection of rubber clones. Some seedlings often have their bark almost entirely filled with stone cells (high stone cell formers). In such seedlings the ratio of 'hard' and 'soft' bark is always unfavourable to high latex production; as a result selection is directed against high formation of stone cells. In addition to the heritability of the trait of high level of stone cell formation, unfavourable environmental factors such as prolonged drought and canopy diseases stimulate the formation of stone cells.

The other structure of the latex producing system worth mentioning is the 'sieve tube system'. Sieve tubes are found in the soft bark near the latex vessel cylinders. The concentration of sieve tubes is always higher near the cambium than in other parts of the soft bark. As the name implies, the function of the sieve tubes is to sieve the latex. It has also been indicated that the sieve tubes play some role in wound healing reactions.

11.2.1.3 Renewed bark

Tapping of the rubber tree is a controlled and scientific wounding and destruction of the bark. As with other latex producing plants, the rubber tree has the ability to heal the wound. After healing, such a bark is known as a 'renewed bark'. Regenerated bark shows three distinct differences from the virgin bark. These are as follows:

1. The phelloderm of the renewed bark is usually very thick (composed of several layers of cells while the phelloderm of the virgin bark is usually thin (composed of only few cell layers). Several rows of the wound phelloderm under the cork cambium contain cell saps which are coloured red by anthocyanin deposits and the red cells are persistent in the regenerated bark.
2. The number of latex vessel cylinders which are formed after the bark has been tapped and regenerated under normal growth conditions is usually higher than in the virgin bark.
3. In renewed bark, the hard bark usually remains thin with a relatively low number of stone cells when compared with those of the virgin bark.

11.2.1.4 Latex formation and flow

The commercial rubber latex is an excretion. The composition of the latex varies (see Table 31) depending on the organ of the plant from which the latex has been extracted – leaf petioles, trunk bark and branch bark. The composition

Table 31 Approximate composition of the latex of *Hevea braziliensis*

Substrate	Percentage
Rubber	20 to 60%
Ash	0.3 to 0.7%
Proteins	1 to 2%
Resins/lipids	2%
Quebrachites	1 to 2%
Lecithins	0.3%
Water (remainder)	33 to 75%

of the latex also varies with the age of the different organs. Generally, the percentage of rubber latex or the volume of latex decreases from the trunk towards the branches and leaves.

Latex flow is an interesting phenomenon. The velocity of latex flow is variable, being lowest in the outer hard bark and in the extreme inner soft bark. It is highest in the middle of the soft bark. The amount of latex flowing out after tapping is directly correlated with the osmotic condition of the latex vessels and of the parenchyma cells which surround the latex vessels. Latex flow is also closely linked with the phenomenon known as **wintering** in rubber. Wintering is the shedding of leaves which usually occurs in rubber around the middle to the end of the dry season in West Africa. Latex flow is high at the commencement of wintering while it is low during the period of new leaf formation.

A number of characteristics of latex flow from rubber trees are as follows:

1. Latex flow from virgin bark of different ages at the same height of tapping panel of trees of the same clone is generally practically the same.
2. Latex flow intensity shows an increase which is roughly parallel to the increasing age of the bark.
3. The amount of latex yield by virgin bark at different heights on the tapping panel varies considerably. It could be as high as the ratio of 1 to 3 over a short distance of about 60 cm.
4. The first renewed (regenerated) bark is known to yield considerably more latex than virgin bark at the same height of tapping panel.

11.2.2 The Environment of Rubber

Hevea braziliensis is a native of the Upper Amazon Basin of Latin America. In its natural home it forms part of the middle storey of the tropical forest. Through agricultural activities, *H. braziliensis* is now cultivated in tropical and subtropical parts of the world where rainfall, temperature and soil permit. The main environmental factors determining the cultivation of rubber are high temperature, adequate moisture and deep fertile soil. Although *H. braziliensis* could grow under moisture stress, it does not produce economic yields under such conditions and, furthermore, it runs into seed production early in life. The commercial product of rubber, i.e. the latex, is water based. Therefore, an adequate moisture supply is indispensable for good vegetative growth and latex production in rubber. Rubber grows best where the rainfall is at least 1750 mm per annum and is evenly distuributed throughout the year.

The temperature range under which rubber performs best is 24 °C to 35 °C provided there is adequate moisture. Rubber is susceptible to wide diurnal temperature variations and, therefore, such areas should be avoided for growing rubber. The latex flow in the latex vessels is a function of cell turgidity and this is related to the environment. Latex flow is known to be higher under high relative humidity.

Soils for the cultivation of rubber should be deep, free from iron concertions, high in organic matter content, well drained and fertile. The land should be as level as possible. Steep slopes should be avoided for the cultivation of rubber. Rubber performs best from sea level up to an altitude of 300 m. Above 300 m altitude, conditions become marginal altitude-wise for the cultivation of rubber. Rubber is a lowland crop, and in this respect it is very similar to cacao.

Rubber is very susceptible to strong winds. Rubber trees are tender and soft. The branches are easily broken by wind. Tracks of strong winds should be avoided for the cultivation of rubber. Although rubber could grow vegetatively in areas subject to periodic desiccating harmattan winds, such rubber trees are known not to give economic production.

Most parts of West Africa are subject to tropical rain storms. Damage through tropical storms is minimized by the use of windbreaks or shelter belts for rubber planting. It must, however, be noted that regardless of the planting of windbreaks, tropical rain storms constitute a menace to rubber plantations in West Africa. Rubber breeders should direct breeding efforts towards the production of dwarf and high yielding rubber varieties as an insurance against tropical rain storm damage to rubber plantations.

11.2.3 Agronomy of Rubber

11.2.3.1 Selection of the site

The selection of a site for planting rubber is determined primarily by soil quality, rainfall patterns, altitude and temperature ranges. The topography of the land is also very important. Steeply sloping or badly dissected lands make operations (transportation, movement of workers especially tappers) more difficult and expensive on the plantations.

Other non-environmental factors which must be taken into consideration when a site is being selected for planting rubber are as follows:

1. Accessibility (good roads, rail lines, water or air transport).
2. Adequate supply of the right type of labour as rubber tapping is a highly skilled and specialized operation.

3. Pipe-borne water from farm dams, bore holes or perennial streams.
4. Electricity supply for processing.
5. Market for disposal of produce.

11.2.3.2 Propagation

11.2.3.2a Seed propagation

H. braziliensis is predominantly cross-pollinated and in rubber breeding it is a standard practice to make crosses by hand. In commercial seed production, seeds are collected from selected trees. Trees are selected on the basis of the records of their performance – yield, growth, freedom from diseases and pests. Commercial seeds are open pollinated. To collect seeds, the surroundings of the selected trees are cleared of all weeds and cover-crop, and in areas where animal damage to fallen seeds is anticipated, the area surrounding the selected trees is fenced with wire netting. During the seed season, all seeds falling within the prepared area are collected daily for planting in the nursery. Rubber seeds lose viability rapidly on dropping from the mother trees. Therefore seeds should be planted immediately on collection. To ensure that viable seeds are sown, viability tests should be carried out on seed samples at regular intervals. To assist early germination, the micropylar cap of the seed should be removed. Seeds thus prepared can be sown either in seedbeds or in seed boxes. Germination is veriable and it can take from 5 to 25 days.

Newly germinated seeds are removed daily from the germinating medium and planted either in well prepared seedbeds at a spacing of 30 × 30 cm or in polybags which have been filled with topsoil. Planting in polybags is at the rate of one seedling per bag.

The following management practices are of importance in a rubber nursery:

1. *Shade* Rubber seedlings require shade during the first half of their life in the nursery period. After that, no shade is necessary. Removal of shade from the seedlings should be gradually carried out in order to avoid a shock to the seedlings.

2. *Weeding* Hand weeding is the standard practice. As much as possible, the use of herbicides should be avoided in rubber nurseries.

3. *Mulching* Seedlings should be mulched during the late rains and just before the commencement of the dry season.

4. *Pests and disease control* Seedlings should be protected against pests and diseases. The best methods of protection in order of importance are: (*a*) adequate nursery sanitation; (*b*) use of chemicals; (*c*) use of biological control available; and (*d*) use of tolerant and/or resistant varieties.

5. *Watering/irrigation* Excessive water supply is as undesirable as inadequate supply of water. Good growth depends on the availability of adequate moisture and nutrients to the seedlings.

6. *Pruning* Seedling growth in rubber is normally monopodial. In some cases, especially where the shoot apex has been damaged, seedlings tend to develop sympodial growth too early. In such cases, intensive pruning is recommended to develop the trunk of the trees. Seedlings under normal conditions of growth will be ready for transplanting into the field when they have grown in the nursery for 12 to 15 months. If they are raised for use as root stocks, they will be ready for budding (patch budding) around the same time or shortly before.

11.2.3.2b Vegetative propagation

The most commonly used method for the vegetative propagation of rubber is budding. Desirable root stocks are raised as seedlings, which are patch budded at the age of 12 to 15 months. The successfully budded materials are stumped at 1 to $1\frac{1}{2}$ months after budding which is about two weeks after the removal of the wrapping tape. Stumping is usually carried out with a pruning saw at a height of about 3 to 4 cm above the bud union. The cut surface is painted over. Transplanting into the field can be done any time from two weeks after stumping. The height at stumping is decided by the method of tapping to be adopted. Generally for a one panel tapping system stumping height is at 2.5 m while it is 3.25 to 3.5 m in cases where a two panel tapping system is to be adopted.

The normal age at stumping is 12 to 15 months. In some cases, stumping can be deferred till the stocks are 2.5 to 4 years old. Some advantages have been claimed for late stumping. Such advantages include: (*a*) higher survival rate of stumps after transplanting: (*b*) earlier plant maturity for tapping; and (*c*) a longer period of intensive care in the nursery for root stock stumps.

Although the most usual and common method of vegetative propagation of rubber is patch budding, it could also be propagated by cuttings or by green budding.

1. *Cutting* This is sometimes used for experimental materials. Efforts are being made in the Far East (Malaya) to root rubber cuttings as a method of vegetative propagation of rubber for commercial plantings.

2. *Green budding* This is the same as patch budding but root stocks are budded when they are three or four leaf-stage seedlings. This method of budding was developed at the Araromi Rubber Estate, Ondo State, Nigeria.

11.2.3.3 Transplanting into the field

As the rubber tree does not require shade, the normal method of land preparation for field planting is clear-felling. After clear-felling all trash is removed or burnt and remaining stumps and roots are removed as far as resources permit. Removal of stumps and roots tends to minimize the risk of the outbreak of root diseases later in the life of the trees.

Cover crops are established in the normal manner: the field is blocked; the blocks are lined out at the appropriate spacing adopted, and the planting holes are dug. The normal size of the planting hole is 60 × 60 × 60 cm except when large stumps are transplanted. In this case the hole is dug large enough to contain the stumps.

Although in the past rubber seeds were planted directly in the field through food farms, it has nowadays become standard practice to raise rubber seedlings in a nursery either for transplanting into the field as seedlings or for use as root stocks.

Rubber can be spaced either in a rectangular or a square pattern, usually at 6 to 7 m. Experience has shown that **avenue planting** has facilitated plantation operations and plant growth. Avenues can be two to four rows with a space of 5 to 7 m between avenues.

11.2.3.4 Maintenance of rubber farms

The major post-planting operations that must be carried out in a rubber plantation are as follows:

1. *Weeding* Weed control is a must in rubber plantations. Where a cover crop is well established, weeding may be confined to the exposed 50 cm circular patch around each plant. It is not advisable to use weedkillers in plantations of young rubber. More work is still needed to elucidate the residual effect of herbicides on latex formation and flow.

2. *Establishment and maintenance of ground cover* One school of thought held that it is preferable to establish legume ground cover after rubber has been planted into the field. This view develops from the recognition of the fact that considerable damage is usually done to cover crops when they are established prior to blocking, lining out, holing and transplanting. Another school of thought, although appreciative of the damage done to cover crops, held that cover crops should be established immediately after clear-felling provided there is adequate moisture. In this case the amount of soil erosion is considerably reduced. The early establishment of the ground cover crop seems advisable!

3. *Mulching* This must be carried out just before the end of the late rains.

4. *Removal of undecomposed mulch* This must be carried out during early rains.

5. *Watering/irrigation* As necessary.

6. *Pruning* To shape the trees, and removal of laterals and unwanted suckers.

7. *Application of recommended fertilizers* This must be carried out at the correct doses and at the appropriate times. Where available rubber is known to benefit from the application of well-rotted organic manures such as farmyard manure or compost.

8. *Regular control of pests and diseases* Root diseases are likely to be a persistent danger in new plantations and they must be kept in check as early as they are detected.

The general growth of rubber is indicative of its productivity. Growth is a function of vigour and, in rubber, the major commercial product comes not from the seeds but from the stem. Therefore, the vigour of growth of the stem which can be followed through girth measurements is a good indicator of productivity. Tappable age is also determined in rubber by girth measurements. There are two growth phases in rubber:

1. Juvenile growth phase, when growth is monopodial and orthotropic.
2. Mature growth phase, when growth is mainly sympodial and plagiotropic.

In budded material, when budwood is taken from plants which are still at the juvenile growth phase, the scions exhibit juvenile type (JT) growth for some time before assuming the mature type (MT) growth. MT scions do not exhibit JT growth type, and they become tappable much earlier. It is also of common occurrence for JT buddings to exhibit intermediate girth sizes between seedlings and MT buddings.

In plantations, it is common practice to commence girth measurement as from the fourth year of transplanting, to determine when the trees are ready for tapping. The normal practice is to open up the trees for tapping when they have attained a girth of 45 cm at a height of 100 cm from the root collar in seedlings or from the bud union in budded material.

11.2.4 Harvesting of Rubber

The main economic product from the rubber tree is the latex. The latex is obtained by tapping the trunk of the tree. Harvesting of rubber is a carefully controlled and systematic wounding of the bark of the rubber tree, and this is known as tapping.

The time of tapping rubber is very important. Latex production is highest in the early hours of the morning. The later during the day tapping is carried out, the less the amount of latex obtained. Extensive studies (Dijkman, 1951) have shown that latex flow begins to decrease around 9.00 to 9.30 a.m. Tapping operations which are aimed at obtaining optimum production from the trees should be finished before 9.30 to 10.00 a.m.

Latex flows out as a result of turgor pressure in the latex vessels and the pressure exerted by the adjacent parenchyma tissues. These pressures are highest during the morning. This phenomenon has been explained by the fact that air humidity is at its maximum in the mornings while transpirational pull is at its lowest. As transpirational pull increases from the morning to mid-afternoon, the turgor pressure of the cells is lowered.

11.2.4.1 Rubber tapping systems

11.2.4.1a The Brazilian method

The first people to tap *Hevea braziliensis* were the Red Indians, followed by the Brazilians. The earliest method of tapping was by extracting latex by daily cuts of 2.5 to 5.0 cm long along the stem of the tree for a period of 100 to 150 days per year. The trees were allowed to rest for the remaining part of the year (usually the rainy season). This is known as the Brazilian method. The defects of this method consist of: (*a*) too deep wounds; (*b*) development of bumpy scars on the bowl; (*c*) rapid consumption of the tappable region of the bowl; and (*d*) excessive contamination of latex by bark fragments. With these defects, the Brazilian method shortens the economic life of the trees.

11.2.4.1b The herringbone system

The herringbone system of tapping was evolved in 1897 by Ridley (Dijkman, 1951). In this system, oblique cuts are made in the bark with all the cuts converging into a central vertical cut. The collecting cup is put at the bottom of the vertical cut. The latex flows from the oblique lateral cuts to the central cut which leads to the collecting cup. The depth of cut is carefully adjusted so as not to injure the cambium which underlies the soft bark. Usually the depth of cut is about 4.0 to 4.5 mm. The Ridley herringbone system has been the only method for tapping Lagos rubber, *Ficus elastica*.

11.2.4.1c The panel tapping system

The panel tapping system evolved as a result of the studies of the anatomy and physiology of the rubber tree. The aim of tapping is to open the latex vessels to enable the latex to flow out. On opening the vessels, latex flows out rapidly at first, then gradually slows down, the pressure decreases and when the vessels are almost empty the latex coagulates, thus plugging the opened ends of the latex vessels. To resume latex flow, the coagulated latex must be removed; this material is known as 'scrap'. The removal of the scrap is followed by removing a bark strip of approximately the same width as that of the coagulant plug from the top ends of the latex vessels. This is the basis for the reopening of the tapping cut.

The depth of the tapping cut should not exceed the innermost depth of the soft bark. Under no circumstances should the cut go as deep as the soft bark/cambium region.

Tapping in the panel system consists of the removal at regular intervals of thin shavings of bark from the surface of a groove made into the bark to a depth of about 1.0–1.5 mm away from the cambium.

The angle of slope of the cut is also very important. The latex vessels in the bark are inclined at an angle of about 3.5° to 3.8° to the right. Therefore, a cut from high left to low right cuts through a larger number of latex vessels per unit of length of tapping cut than a cut from right to low left. Experimental data have shown that cuts from high left to low right produce 8 per cent more latex at an angle of 25° than horizontal cuts. However, cuts from high left to low right produce 14 per cent more latex at an angle of 45°.

Tappers, on the basis of experimental data and convenience of tapping, have internationally adopted tapping cuts of 25° slope to the horizontal for seedling trees and 30° slope for budded trees. Normally the annual bark consumption is roughly 30 cm. First, young trees are opened for two years at a height of 65 to 70 cm above ground level or from the root collar. In any case, the lowest end of the tapping panel should be 12.5 to 15 cm above the highest point of the bud union. Before tapping the positions of panels on the trees are marked either using paint or by making superficial scratches on the trees with the tapping knife. The sequence of the operations is an follows:

1. A vertical front line is marked down to the ground level along a ruler held against the tree up to a height 100 cm.
2. The height of the tapping cut is marked on the vertical front line by a short horizontal scratch extending to the right.
3. The horizontal girth of the tree is measured with a string at the tapping height (65 or 70 cm).
4. Depending on the tapping system to be adopted, the girth is divided into two, three or four equal parts.
5. Another vertical groove marking is made on the bark to indicate the rear side of the panel.
6. The tapping cut is then marked at an angle of 30° with the horizontal or 60° with the rear side vertical marking. This is the first tapping cut mark.
7. Parallel to the first tapping cut mark, two-month-bark consumption lines are marked into the bark.
8. The spout is fixed 15–20 cm under the tapping cut mark across the bottom of the front line of the panel.

These operations are carried out by the tappers, working together in groups of two, with a tapping supervisor overseeing the entire work.

In tapping of rubber trees the amount of tapping is always expressed as a fraction of the circumference, the regularity of tapping, the length of the tapping cut, the number, the tapping period and circle, the changeover systems, the number of successive panels, periodicity and relative tapping intensity are all denoted by internationally accepted notations.

The basis of the International Notation for tapping systems in rubber is a statement of the amount of tapping as a fraction of daily tapping on a full circumference or girth of the tree. Notations are always expressed as fractions in which the **numerator** consisting of capital letters denotes the type of cut, and the **denominator** is the reciprocal of the fraction expressing the horizontal length of each cut in terms of full circumference.

1. S/1 denotes a spiral cut on full circumference.
2. S/2 denotes a sprial cut on one-half of the circumference.
3. S/3 denotes a spiral cut on one-third of the circumference.
4. S/4 denotes a spiral cut on one-quarter of the circumference.
5. 2S/2 denotes two half circumference spiral cuts.
6. V/2 denotes a V cut on half circumference.
7. L/1 denotes extended V (left arm of a half V cut extended upwards to full circumference).
8. C/2 denotes a half circumference cut (horizontal).

Tapping period or periodicity of tapping is expressed as a fraction of a cycle of tapping plus rest, reckoned in days, weeks, months or years. In this case, the numerators are in small letters (d = day; w = week; m = month and y = year) and the denominators are numerals:

1. d/2 denotes tapping on alternating days.
2. d/3 denotes tapping once in three days.
3. d/2, 6m/9 denote alternate daily tapping for six months followed by three months of rest.
4. d/1, m/2 denote daily tapping for one month out of two followed by one month of rest.

The number of successive panel tapping periods and cycles of changeover of panels are expressed by number of successive panels, 1, 2, 3 or 4 and so on multiplied by the fraction indicating the period of tapping plus the resting period between each change of panel over the cycle of panel rotation, e.g.

1. 2 × 6m/12 denotes two panels, each tapped for six months alternately, making a total panel cycle of 12 months.
2. 3 × 6m/18 denotes three panels, each tapped for six months tapped once in the three days, making a total panel cycle of 18 months.

The standard intensity of tapping rubber is 100 per cent which means that the trees are just tapped to normal capacity, taken to be S/2 d/2. Intensity of tapping is calculated by multiplying the notation formula indicating the tapped part of the circumference and the tapping period by a factor 400. Therefore,

$$\text{S/2 d/2} \times \frac{400}{1} = 100\% \text{ tapping intensity}$$

This intensity of tapping is subject to alterations. If, for example, Sundays are accounted for the intensity will become

$$\text{S/2 d/2} \times \frac{400}{1} \times \frac{6}{7} = 86\% \text{ tapping intensity.}$$

The intensity of tapping can, of course, also be altered by changes in the periodicity of tapping, the depth of resting periods, the length of the cuts, and so on.

11.2.4.1d The double panel system

When rubber trees are growing under extremely favourable conditions and the market situation is brisk, it at times pays to adopt a two panel tapping system so as to maximize latex production without endangering the life and productivity of the trees. A two panel system is recommended for use with **healthy** and **vigorous** trees only. It could be:

$$\text{2 S/2 d/3} = 133.3\%$$

or

$$\text{2 S/1 d/3 6m/9} = 177.8\%.$$

With healthy trees, double panel tapping with two cuts, one above the other, either on one side of the tree or opposite sides is physiologically normal, because (*a*) the reaction of one cut is known to be independent of the other (Dijkman, 1951), (*b*) the latex flow depends on the genetic capacity of the tree, and because (*c*) the area of latex extraction and regeneration is extended.

The following precautions should be taken when two panel systems of tapping are adopted.

1. The higher panel should be at least 240 to 250 cm above the ground level, i.e. about 150 cm above the lower panel. The spacing between the two panels should not be less than 125 cm but could be up to 180 cm.
2. Double panel tapping should only be used with trees that are over 10 years of age and are in good health.
3. Appearance of brown bark (BB) is a confirmation of overtapping, and a rest period should be introduced, i.e. the intensity of tapping should be reduced.

11.2.4.1e Periodic systems

Rubber tapping is a controlled and systematic wounding of the bark. The renewal of the wounded bark is a cambial activity which takes time. Therefore, it is necessary to introduce rest periods to reduce tapping intensity so that bark consumption does not outstrip bark renewal. This can also be accommodated into the International Notation, e.g.

1. C/2 d/2 6m/9 = 67/, i.e. half circumference cut, tapped alternate days for six months out of nine, and rested for three months. In this case, the tapping

area is divided into three A, B and C. Areas A and B are tapped for three months while C is rested, etc.
2. C/2 d/2 9m/12 = 75%, i.e. the tapping area is divided into four (A, B, C and D). Each block is rested for three months in rotation and tapped for nine months.

Experience shows that S/2 d/2 6m/9 = 67% generally over a long period gives higher yields per hectare than longer periods such as 10m/15 or 12m/18.

11.2.4.2 Practical tapping hints

The tools and materials required for tapping are the tapping knife, the spout, collecting cup, cup hanger, collecting buckets, churns, collecting tanks, anti-coagulants (sodium sulphate, ammonia). Cleanliness is absolutely essential in rubber tapping. In addition, taping and arrangements for tapping are easier on flat lands. On hilly land tapping is difficult. When tapping has to be done on sloping land, paths should be made along the contour to facilitate the carrying of latex to the collecting centre by tappers. The identification of a tapping task is a problem on any rubber plantation. A number of factors including the age, health and variety of trees, the experience of tappers, the topography of the land, influences the possible size of a tapping task. On average, a tapper is expected to tap 300 trees per day.

Each morning of tapping, each tapper goes to his task with his tapping knife, container for scrap rubber, and his bottle of anticoagulant. On reaching each tree, he removes the scrap rubber, treats the collecting cup with anticoagulant, replaces it and then taps the tree by removing a thin strip of bark from the top end of the panel bark.

On collecting the latex, the latex brought in by each tapper is weighed and the dry matter content is determined with a hydrometer graduated to measure latex of about 35 per cent d.r.c. (dry rubber content). The average d.r.c. of latex from a mixed plantation of differently aged trees is 30 to 40 per cent. After tapping has been carried out for about two weeks, the daily task is worked out based on the average performance of the best 10 per cent of the tappers.

On large plantations it is essential to set up latex collecting centres at convenient points. The floors of such latex collecting centres are concreted and roofed with iron sheets. It is there that latex is received from tappers, weighed, d.r.c. checked, strained, bulked and treated with more anticoagulants prior to transportation to the processing plant.

11.2.4.3 Stimulation of latex flow

Under certain circumstances, such as a brisk market situation, the need to rehabilitate plantations with declining yield of latex as a result of tapping cuts overlapping barks of different ages, or on trees that are over 15 years old and are in excellent condition of growth, latex flow (latex yield) can be stimulated.

Stimulation can be achieved by lightly scraping the outer layer of bark from just below the tapping cut or by removal of ageing leaves from the lower branches. The use of stimulating compounds is another possibility.

A number of chemicals have been used to stimulate latex flow in rubber. 'Newbark' – a compound containing a sulphate of iron and potassium permanganate. A 15 cm strip is scraped round the trunk just below the tapping cut and the compound is painted over the scraped surface. Other compounds that have been used to stimulate latex flow are nitrate of soda, wood ash, cattle manure, vegetable and mineral oils, palm oil which increases latex flow and accelerates bark renewal in young trees when applied just above the tapping cut, synthetic hormones in oil carriers such as 2, 4-D and the butyl ester of 2, 4, 5-T.

11.2.5 Processing of Rubber

11.2.5.1 Preservation of latex

The normal practice is to process latex immediately on arrival at the processing plant. However, circumstances arise when latex has to be stored. The essential operations for storing latex are as follows:

1. The latex on arrival at the processing plant is bulked and strained.
2. Gaseous ammonia, at the rate of 22 ml of strong ammonia solution or 31 ml of 20 per cent ammonia solution per litre of latex, is added. Thus treated, the latex can be stored or exported liquid without any fear of coagulation.

11.2.5.2 Processing methods

The essential operations in the processing of latex are the following:

1. On arrival at the processing plant the latex is diluted with a small volume of clean water and strained into the bulking tank. The main reason for this dilution is to aid effective straining.
2. The bulked latex is diluted, under continuous stirring, with more water to obtain a d.r.c. of 130 to 150 g per litre of latex for sheet rubber, 200 g for crepe, and 150 to 200 g per litre for crump. Latex rubber can be processed into crump rubber, crepe rubber and/or sheet rubber. The international market for rubber has always favoured crump rubber for which higher prices are paid.

11.2.5.2a Processing of latex to crump rubber

Many processes have been devised for processing latex to crump or powdered form of rubber. In the **Pulvatex** (Stam) process, latex to which an appropriate volume of a protective colloid, e.g. diammonium phosphate, has been added is sprayed by means of a centrifugal device into a heated air stream. The spray droplets dry up rapidly and dry crump or powdered rubber is obtained. The choice of the colloid is very important. Colloids such as dextrin will produce crumps which will mass together on cooking, while diammonium phosphate will eliminate such massing of crumbs.

In the **Mealorub** method, fresh latex is treated with stabilizers together with sulphur, zinc oxide and other vulcanizing substances. Such latex is stored for a long time, or heated to 80 °C for $2\frac{1}{2}$ hours, until the adherence of the coagulum disappears. The latex is then coagulated into a crumbly mass, dewatered, mechanically disintegrated and air dried.

In a third (**Heveatex**) method, latex which has been treated with a hydrophilic stabilizer such as casein is flocculated by an agent (e.g. zinc chloride) to render the stabilizer insoluble. The floss is separated, dewatered and the crumbly cake (mass) is granulated. For proper granulation, a water-insoluble power such as zinc stearate should be added.

In the **nitrite crumb** method, fresh latex is treated with sodium nitrite and then coagulated with acid. The coagulum is broken up by continuous stirring into a crumbly cake which is dewatered, disintegrated and dried.

11.2.5.2b Processing of latex into sheet rubber

Until recently, this has been the most popular method of processing latex. However, this has changed in favour of processing into crumb. After the latex has been appropriately diluted, it is strained into a bulking and settling tank. The diluted and strained latex is allowed to settle for 15 to 20 minutes, after which it is let out into the coagulating tank. The coagulating tanks are specially constructed preferably with aluminium alloys. The partitions are made of the same material and a standard sized tank of 3 m long by 90 cm wide is designed to contain 90 to 95 partitions. The sizes of coagulating tanks most commonly used on estates are $3 \times 0.9 \times 4.0$ m and with a capacity of 13 m^3 or $3 \times 0.9 \times 5.4$ m and a capacity of 14.6 m^3. In the olden days, the tanks were designed to produce non-continuous sheet, i.e. 80 or 90 separate slabs of coagulum from each tank, but modern tanks are constructed to produce a single continuous sheet from each tank.

Coagulation is carried out by the use of coagulants. The most commonly used coagulants are formic or acetic acid, diluted with water according to the instructions on the label for coagulating latex. When acetic acid is unavailable, sulphuric acid, coconut water, coconut toddy, palm wine (undiluted) can be used as coagulants. On the addition of the coagulant to the latex, the tank should be agitated to ensure a uniform coagulation of latex.

After coagulation, the sheet of coagulum is machined, under a continuous flow of clean water, into thin sheets for drying. The more plastic the coagulum the better it will form sheets. Coagulum is tender and plastic and could be easily destroyed by careless handling; therefore handling of coagulum should be reduced to the barest minimum. The coagulum should be conveyed from the coagulation tank to the sheeting battery by a chute, having been lifted or floated into the chute. Before re-using the coagulation tank, it must be thoroughly cleaned with water.

The rubber sheets are then passed through mills which consist of rollers (four to six pairs) arranged in a linear order with the final pair of rollers being grooved so as to produce the criss-cross ribbing which gives the smoke-dried product its name: ribbed smoked sheet. As the sheet passes through the last pair of rollers, it is cut into convenient but regular sizes for smoke drying. The sheets are carefully and cleanly washed and transferred in trolleys to the smoke houses for drying.

Smoke houses are constructed on the principle that sheets should not be carried from the ground level, but should be transported into the smoke houses in trolleys pushed along rails. Lifting of sheets is avoided, and handling is limited to the loading and the off-loading of trolleys.

The 'single subur' smoke house is built of four chambers, all served by one furnace, placed centrally at one end of the building, below ground level and with a central main flue running the whole length of the building, with smoke outlets and branch flues serving each chamber. The building is provided with entrance and outlet doors. Inspection windows are also provided. By manipulation of flue outlets and ventilators temperatures in the first chamber should be at 37 °C for three hours, rising later to 43 °C to 48 °C. In the second chamber temperatures are to be 49 °C to 52 °C; in the third chamber 54 °C to 60 °C; and in the fourth chamber 60 °C to 66 °C.

The 'double subur' smoke-house consists of two 'single suburs' joined by a central verandah. The double subur is used on very large and productive plantations.

Another type of smoke house is the tunnel type. These are designed for specific quantities of produce. Generally, there are three standard types commonly known as types A, B and C which are designed to handle 450, 900 or 1800 kg dry sheets per day. Tunnel-type smoke houses are tunnel-like in construction. Usually type A is about 11 m long, 2.4 m wide inside with doors at both ends and capable of holding four to six trucks on a single rail track along the centre. The main flue enters from one side, with branch flues with outlets under the centre of each truck. Tunnel-type smoke houses are nowadays preferred to subur houses on the grounds that tunnel-type houses have a lower initial cost, greater simplicity and are easier to operate. They incorporate the main advantages of the subur type such as continuous smoking, avoidance of lifting or carrying of sheets and low fuel consumption. They also have the following advantages over the subur type:

1. No turntables are required.
2. Excavation is reduced to the minimum.
3. Furnace efficiency is enhanced.
4. Trucks are loaded from the side (a distinct advantage).
5. Flues are more easily accessible.

6 The trucks and floors facilitate rapid drainage of dripping water from the floors.

Type B is basically the same as type A but the dimensions are generally 16 m in length and 3.3 m wide inside; type C is a double B type, either arranged in line or parallel, depending on the location.

Under native conditions sheet rubber is prepared by coagulating small quantities of latex (about 225 ml at a time) in boiling water and immediately flattening the coagulum into a thin sheet of rubber with a wooden roller on a table. This flattening is done by hand. Most of the rubber exported from West Africa during the Second World War was processed in this way.

Smoked sheet rubber is graded with the following conditions:

1. Wet, bleached and virgin rubber is not acceptable for sale.
2. Skim rubber made of skim latex should not be used in whole or in part in the production of any grade of rubber meant for the international market.
3. Copper and manganese contents should not exceed 8 p.p.m. and 10 p.p.m. respectively.

Rubber grades as recognized by International Codes are as follows:

1. No. IX RSS = Superior Quality Rubbed Smoked Sheets.
2. No. 1 RSS = Standard Quality Rubbed Smoked Sheets.
3. No. 2 RSS = Good Fair Average Quality Rubbed Smoked Sheets.
4. No. 3 RSS = Fair Average Quality Rubbed Smoked Sheets.
5. No. 4 RSS = Low Fair Average Quality Rubbed Smoked Sheets.
6. No. 5 RSS = Inferior Fair Average Quality Rubbed Smoked Sheets.

RSS cuttings are trimmings which are obtained from smoked sheets during packing. They are generally of two grades – A and B.

Grade A – Permitted to contain slight traces of fine non-gritty carbon dust, heavily smoked, slight rust and with up to 5 per cent mould.

Grade B – Permitted to contain up to 10 per cent speck and/ or bark/mould, bark clippings with no other extraneous matter.

Cuttings are usually wrapped in RSS of not lower quality than No. 4 RSS.

Sheet rubber is usually packed in bales with maximum weight of 114 kg net per 0.45 m^3 and a minimum weight of 112 kg net. For No. IX RSS, No. 1 RSS and No. 2 RSS, the outside of the bale must be dusted with powder before applying the wrapper sheets. This is necessary to prevent sticking.

11.2.5.2c Other rubber processing methods

Through appropriate processing methods these other important types of marketable rubber products are obtained.

1. *Crepe* Generally, crepe rubber is produced from fresh coagulum of natural rubber latex under carefully controlled conditions. The rubber is creped in thickness corresponding to the standard sample. Crepe rubber is classified mainly according to thickness.

2. *Cyclized rubber* Cyclized rubber is processed mainly in Malaya. In this process, the rubber molecules are changed from a linear to a cyclic structure (cyclization) by treating the latex with sulphuric acid. Cyclized rubber is generally processed for the manufacture of shoe soles, and is a good substitute for synthetic resins.

3. *Superior processing rubber* (*SP rubber*) SP rubber is a form of natural rubber in which a portion of the raw latex is vulcanized before coagulation. SP rubber has superior processing properties which have made it possible to use natural rubber for purposes from which hitherto it had been excluded. Briefly, the process consists of two stages. During the first stage a stable suspension in water is prepared of chemicals necessary for rubber vulcanization. The chemicals concerned are sulphur, zinc oxide, zinc diethyl, dithiocarbonate and mercaptobenzene thiazole. This suspension is then mixed with a dispersing agent. Field latex is strained, ammoniated to 0.3 per cent and mixed with the chemical slurry, which is brought to 82 °C for two hours. During the second stage the vulcanized latex is blended with strained field latex (1:4), stirred and then coagulated. The material is then milled, dried and packed as sheet rubber.

11.2.6 Diseases of Rubber

11.2.6.1 Root diseases

There are three major root diseases of rubber. These are:

1. White root caused by *Fomes lignosus*.
2. Brown root disease caused by *Fomes noxius*.
3. Red root disease caused by *Ganoderma pseudoferreum*.

The aetiology of the three diseases is very similar. The causal organisms are fungal. The rhizomorphs of the spreading mycelia grow along the roots and become firmly attached to them. For *F. lignosus*, the rhizomorphs are typically white and flat at the growing points, for *F. noxius* the rhizomorphs are coloured tawny brown, covered with a surface crust of irregular thickness and become almost black with age. The rhizomorphs of *G. pseudoferreum* form a continuous tough skin all over the surface of the root, usually creamy white at the growing points and turning red as the mycelia become old.

The symptoms of the three root diseases listed above are very similar. The diseases affect the trees by cutting off supplies of nutrients and water to the crown, thus causing water stress and starvation in the trees. Foliage becomes thin, unhealthy and turns brown with the leaves falling off. The branches starting from the growing point begin to die back, latex yield is reduced possibly to nil and eventually the trees die. These symptoms are confirmed by the presence of the mycelia of the causal fungus on the roots.

Farm sanitation is an important control measure. This involves thorough preparation of the land for planting through clear-felling, removal of tree stumps and roots and avoidance of sites with rotting roots when planting the rubber seedlings. It is also important to carry out root inspection of tree collars for infection at regular intervals, followed by treatment of affected trees. Affected roots can, after exposure, be treated with appropriate fungicides such as Captan or Bordeaux mixture.

11.2.6.2 Stem and branch diseases

11.2.6.2a Stem rot

This is caused by *Ustilina zonata*, a wound parasitic fungus and saprophyte of dead rubber wood. The fungus rarely infects undamaged healthy trees. *Ustilina zonata*, unlike the *Fomes* spp. does not form rhizomorphs nor spread through the soil by growing over organic matter. Spread is achieved through contact and spores which may be windborne or dispersed by insects.

Infected trees usually exhibit dry, light-brown to whitish, readily fragmented branches with a network of irregular black lines running through them. Just below the bark, grey-whitish sheets of mycelium usually occur. The fruiting body of *U. zonata* is a flat grey-black structure, attached to the bark only as its point of origin but closely appressed to the bark.

Control of this fungus is through farm sanitation, by cutting off infected wood, the use of resistant clones and treatment with appropriate fungicides.

11.2.6.2b Stinking root rot

Caused by the fungus *Sphaerostilbe repens*, This disease is of rare occurrence when rubber is grown under healthy conditions, where the soil is well drained and the topography gently undulating. The fungus generally attacks the roots with symptoms similar to those of other fungal root diseases of rubber. The smell is not produced by *S. repens* but by the numerous deposing saprophytic fungi and bacteria which associate with *S. repens* under the wet conditions in which it thrives. Control is by removal and burning of infected tissue, by farm sanitation, by adequate drainage and aeration of the plant site and by the use of appropriate fungicides.

11.2.6.2c Pink disease

This is caused by the fungus *Corticium salmonicolor*. It attacks rubber, other fruit trees such as cacao, coffee, tea, kola and many forest trees. Pink disease is usually very severe on rubber trees that are under the age of 10 years. Older trees do not as a rule suffer severe attack. The surface of the bark of affected trees is usually covered with a fine silky white cobweb, this giving rise later to the salmon-pink fungal growths from which the disease derives its name. Control is through farm sanitation, the avoidance of mixed planting of host plants and the use of resistant varieties.

11.2.6.3 Panel diseases

11.2.6.3a Mouldy rot

This is a disease of rubber growing under damp conditions. The causal fungus is *Ceratostomella fimbriata.* The fungus is unable to survive under conditions of low humidity; it has an olive-brown mycelium and three kinds of spores:

1. Colourless thin-walled cylindrical spores which are produced in large quantities on the surface of newly tapped bark which are responsible for the characteristic appearance of the fungus on the affected panel.
2. Olive-brown thick-walled resting spores formed in small quantities.
3. Colourless thin-walled spherical spores produced in black flask-shaped bodies.

The disease is spread mainly by spore-contaminated tapping equipment (knives, collecting cups). The early symptoms of this disease include the appearance of a series of slightly depressed and discoloured spots on the panel, which later darken and become covered with greyish mould. The spots spread and in advanced stages cover the whole tapping cut. Farm sanitation, adequate aeration and the use of clean tapping tools and materials are the best methods of control. Once the panel has been affected, it should be treated with appropriate fungicides.

11.2.6.3b Black thread or black stripe

This is not a very common disease. When it occurs, however, it can be very severe and cause serious losses. The causal organism is a species of *Phytophthora*, the same genus to which the fungus causing black pod disease of cacao belongs.

At the early stages, the disease is characterized by black pencil-like lines in the cortex and wood of infected trees. Infection usually starts at the tapping cut and runs vertically upwards. High humidity favours the growth of the fungus and spread of the disease. Control is through farm sanitation and by pruning of trees and cutting of ground cover to aid free air movement within the plantation. The use of resistant clones may prevent occurrence of this disease but once affected the trees should be treated with appropriate fungicides.

11.2.6.3c Bark bursts or cortical fissures

This is a non-pathological disorder in rubber. It is the physiological split of the bark of rubber trees. This disorder can occur as a result of sudden resumption of vigorous growth after a period of dormancy. Bark bursts are common during the early rains following the main dry season. Internal strains arising in the cortex as a result of wind stress on the trees can also lead to such fissures.

11.2.6.4 Deficiency diseases

As with other plants, the rubber tree responds to the availability and non-availability of nutrients in the soil. The most important nutrients for rubber are nitrogen, phosphorus, potassium, magnesium, calcium, sulphur iron and manganese. Some of the characteristic symptoms of the deficiencies of these nutrients are as follows:

1. *Nitrogen deficiency* Stunted growth: small, yellow or yellowish green leaves; necrotic leaf tips and premature leaf fall, stoppage of growth or, where growth continues, new shoots are short and thin; unthrifty growth.

2. *Phosphorus deficiency* Weak, thin and spindly growth; leaves dull, dark green at first later turning olive-green; upper leaf surface yellowish brown or bronze while the underside is purplish; in severe cases leaves become scorched.

3. *Potassium deficiency* Leaves pale bright yellow with orange speckling; marginal necrosis known as rim scorch and premature leaf fall.

4. *Magnesium deficiency* Intense yellowing of interveinal spaces, appearing first on the lower leaves, followed by necrotic patches on the leaf veins.

5. *Calcium deficiency* Marginal necrosis of younger leaves; roundish necrotic discs dropping out of leaf lamina, thus giving the leaves a ragged look. This is a very distinct feature of calcium deficiency in rubber.

6. *Sulphur deficiency* Stunted growth, yellowish and sometimes marbled leaves.

7. *Iron deficiency* Acute interveinal chlorosis of the young leaves; leaf size greatly reduced; premature leaf fall; extensive die-back of shoots.

8. *Manganese deficiency* Symptoms similar to those of magnesium deficiency. Yellowing is usually dull and there are no necrotic patches.

All these deficiencies and their symptoms can be easily cured by the application of appropriate fertilizers.

11.2.7 Pests of Rubber

The rubber tree is relatively free from the attack of pests in West Africa. The major pests that have been associated with rubber are as follows:

1. *Termites* These attack roots and stems especially of unhealthy trees. Control is achieved by locating the nests and treating with Rogor 40.

2. *Cockchafer grubs* These are beetles belonging to the families of the Melolonthidae and Rutelidae in the order of the *Coleoptera*. These beetles are usually large, horned thickly built and hardy. They are powerful flyers. They attack the leaves and shoots of the rubber tree. They can be controlled by:

(a) hand picking and killing;
(b) spraying with insecticides or the use of soil fungicides;
(c) farm sanitation.

3. *Caterpillars* Many species of caterpillars attack leaves and bark of rubber. Usually they do not attain pest status. If occasionally caterpillars become a pest they can be controlled by spraying with an appropriate insecticide. The most reliable method of keeping caterpillars below pest status is to maintain adequate farm sanitation.

Among other pests of rubber are snails, slugs, rodents, domestic animals and bats. Appearance of any of these on the rubber plantation should be discouraged.

11.3 FUTURE DEVELOPMENT OF RUBBER

The world has accepted rubber products as indispensable to the welfare of man and his environment. Barring the normal ups and downs of trade, demand for rubber appears to increase steadily with time. The threat of synthetic rubber (synthetic resins) should not discourage the production of natural rubber.

Chemists will endeavour to eliminate the defects of synthetic rubber but until that is done, natural rubber will hold its place in the world market. Nevertheless, more research work is needed on rubber cultivation, harvesting, processing and latex conservation. The objective should be to reduce the present high cost of producing natural rubber. This can be achieved by breeding high yield trees that are resistant to diseases, pests and environmental stresses. More efficiency is needed in rubber processing plants to minimize wastage of latex. Regardless of any improvement in the production of rubber, there is every likelihood that demand for rubber will outstrip supply very shortly. This underlines the importance of developing the production of rubber even further.

ADDITIONAL READING

Annual Reports of the Rubber Research Institute of Nigeria, 1960 to 1975.
Dijkman, M. J. (1951). *Hevea, Thirty Years of Research in the Far East*. University of Miami Press, Florida.

Edgar, A. T. (1958). *Manual of Rubber Planting* (*Malaya*). Incorporated Society of Planters, Kuala Lumpur, Malaya.

Rand, R. D. (1942). Hevea rubber culture in Latin America. *India Rubber World*, 106–243; 350–356; 461–465.

Rand, R. D. (1945). Hevea rubber culture in Latin America, problems and procedures. In *Plants and Plant Science in Latin America*. F. Veerdoorn (ed.). Waltham, Mass. Chronica Botanica pp. 183–199.

CHAPTER 12

Cashew

The cashew plant, *Anacardium occidentale*, is a native of Central and South America with its main centre of variation in eastern Brazil. Cashew is now grown in many parts of the world where its growth is not limited by cold. The annual world production of cashew nut – the main commercial product of the cashew plant – is about 250,000 tonnes and more than 50 per cent of this production comes from South Asia and East Africa, especially India and Tanzania. Small quantities of cashew nuts are produced in West Africa, in the Mediterranean area, and in other parts of the world The world cashew nut production comes from both wild and cultivated trees. In addition to the nuts, cashew nut shell oil (CNS oil) is obtained from the seed pericarp by steam distillation or extraction with solvents. The cashew apple (Figure 36) is a valuable source of raw material for the manufacture of both soft and alcoholic drinks. The pulp of the apple is an admirable livestock feed. Unprocessed, the nuts of the cashew-nut is very astringent, the roasting of the nuts during processing destroys this astringency.

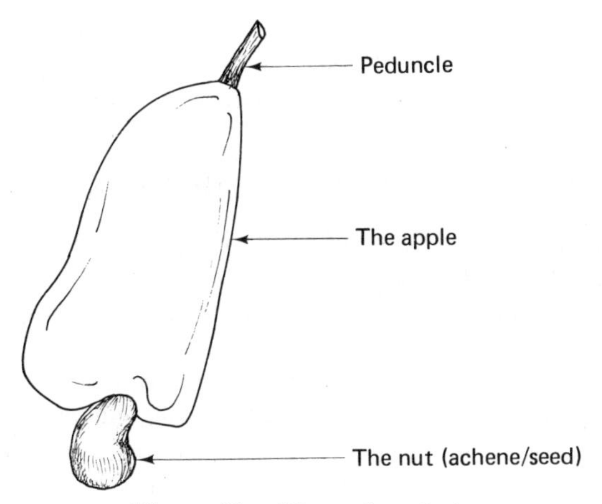

Figure 36. The cashew fruit

The cashew plant crossed the Atlantic Ocean to the old world with the trans-Atlantic trade and was brought to West Africa, to East Africa and to South-East Asia. As a crop plant that is very tolerant of poor soils, low soil moisture content and low rainfall, it has proven to be a good cash crop in areas where other tree crops produced little or nothing.

12.1 TAXONOMY AND BOTANY

Anacardium occidentale is a member of the family Anacardiaceae, the family to which also the mango (*Mangifera indica*) belongs. The cashew tree is of medium size with a spreading canopy often with branches that droop. The mature height is 10 to 11 m. It spreads its canopy very widely. This habit of growth has its implications for the husbandry of cashew. It is dicotyledonous plant with woody trunk, taproot system, with its anchorage roots deeply penetrating into the soil while the feeding roots are generally distributed within the top 50 cm of the soil profile.

Leaves are large, simple, entire, pale-green, relatively thick, alternate, and borne on short swollen petioles. The veins are prominently reticulated.

The inflorescence is either axillary or terminal, the spikes bearing many florets. Male and hermaphrodite flowers occur together in the same loose panicles with about six to seven times as many male flowers as there are hermaphrodite flowers. Flowers are regular, with five green sepals (calyx), five reddish petals (corolla) and ten free stamens. Usually, the filaments of the stamens are equal in length, but occasionally there occur variations in the length of the filaments. The stigma is slightly bulbous. Pollen grains are sticky and large. Pollination is by insects. The ratio of seed set to the number of flowers produced is very low. The fruit is a kidney-shaped achene, with a greyish-coloured pericarp. There is considerable variation in the colour of the seed pericarp and this character is used for varietal classification. The hard pericarp contains the commercially valuable CNS oil (a skin irritant which is used industrially). The cashew apple is either eaten as a dessert, a cooling suck or processed into soft or alcoholic drinks (India and Ghana), when used at all.

12.2 THE ENVIRONMENT OF THE CASHEW PLANT

The cashew plant is adapted to a very wide range of ecological conditions. It is particularly drought tolerant. Only a few economically important tropical tree crops (mango and guava) are approximately as drought tolerant and adapted to a wide range of soil and climatic conditions as the cashew. However, cashew is very intolerant of cold. Cashew is commonly grown on poor soils that will not carry a good crop of cacao, rubber, kola, guava, and even plantains and bananas. Nevertheless, cashew requires deep, well-drained soils (taproot system) free from iron concretions. It will respond favourably to high levels of organic matter and mineral nutrients. Although cashew will grow very well in areas with high rainfall, it is advisable to avoid such areas for its cultivation. Cashew fruit rot is highly favoured by high rainfall and humidity. The best crop of cashew has been reported from areas under annual rainfall regimes ranging from 900 to 1100 mm per annum (India, Tanzania, Nigeria). The rainfall should be evenly distributed over at least 9 to 10 months of the year. Temperature is very important for good growth of cashew. The best crops of cashew have been reported from areas with maximum temperatures of 35 °C, the optimum range for cashew is 24 °C to 29 °C.

Direct insolation and clear or cloudless sky are very advantageous to good growth in cashew. Cashew invariably performs very poorly in areas with excessive overcast skies. Although cashew is drought tolerant, it should not be exposed to excessive winds. In areas with tropical storms, cashew trees suffer considerable damage from wind by mutilation of the canopy. Harmattan winds are injurious to good production in cashew. Cashew will grow well from sea level up to an altitude of 1200 m.

12.3 THE AGRONOMY OF CASHEW

12.3.1 Selection of a Site

Unlike some other tropical tree crops, cashew can be grown under a wide range of ecological conditions. Cashew is reasonably tolerant of low moisture content in the soil while its growth is not adversely affected by high soil moisture content. It will, however, not tolerate waterlogged conditions. The soil must be deep, freely drained and the land reasonably flat.

12.3.2 Propagation

Cashew is mainly propagated by seed. Investigations are in progress on vegetative propagation but these are still at the experimental level.

12.3.2.1 Seed collection

In areas where cashew cultivation is well developed (India, Tanzania) élite seeds are obtained from cashew breeders, for commercial planting. Seeds are collected from mother trees in a way similar to seed collection for rubber trees. Mother trees are identified, the surroundings clean-weeded and seeds which drop on full maturity are collected (during the season) every other day and stored prior to planting or distribution to planters.

12.3.2.2 Storage of cashew seeds

Cashew seeds can be stored under cool dry conditions for about six months without loss of viability. Storage under warm or hot conditions usually results in rapid loss of viability. Viability of cashew seeds can be checked by flotation as for mango seeds. Storage can be in small jute bags in a well-aerated and cool room in racks. Where jute bags are not available, the seeds can be stored in small cane baskets on racks or lightly spread on a dry floor. It is essential that the store is cool and well aerated.

12.3.2.3 Germination of cashew seeds

Cashew seeds possess a very thick coat and because of this the seeds take a considerable time to germinate. Early germination can, however, be induced by the following means:

1. Cracking the sead coat – this is a delicate operation which must be carried out with care to avoid damage to the embryo and the cotyledons.
2. Treatment with dilute sulphuric acid (H_2SO_4).
3. Soaking in water for 24 to 36 hours before sowing in seed boxes.

With any of these treatments, the seeds will germinate in two to three weeks.

On germination, the germinated seeds are planted in filled polybags – one germinated seed per bag.

12.3.2.4 The nursery

Polybags are arranged in the nursery, mulched slightly and regularly watered. No shade is normally required. In very dry areas, light shade may be provided. Disease and pest control measures are regularly carried out during the nursery period.

Where polybags are not available, germinated seeds are planted in nursery beds at a spacing varying from 30 × 30 cm to 45 × 45 cm. The nursery beds are mulched, watered when necessary and disease and pest control measures carried out. Where necessary, light shade may be provided. The nursery period ranges from four to six months at the end of which the seedlings are transplanted into the field when the rainfall is steady. Cashew seeds can also be sown at stake.

12.3.3 Field Planting

12.3.3.1 Field preparation

Cashew plants do not require shade in the field. The desired method of land preparation is, therefore, clear-felling. After land preparation, the plantation is divided into blocks of 4 ha units. Each block is lined out at the spacing chosen.

12.3.3.2 Spacing

The recommended spacings for mature cashew plants are 9 × 9 m, 10 × 10 m, or 12 × 12 m depending on the variety and the recommendations of the seed supplier. It is normal practice to take advantage of dense populations in the early life of a cashew plantation. In this case, and especially with the wide spacings, seedlings are transplanted into the field at 4.5 × 4.5 m, 5 × 5 m or 6 × 6 m and later thinned to 9 × 9 m, 10 × 10 m or 12 × 12 m, after some years of cropping. Some planters, however, prefer the wider spacing from the start and intercrop the plantation during the early years of the establishment of the plantation. The planting holes are 60 × 60 × 60 cm.

12.3.4 Post-planting Field Maintenance Operations

The main post-planting maintenance operation is control of weeds. During the early years ring weeding will suffice. The establishment of a leguminous ground cover helps in weed control, unless intercropping is practised.

Another important operation is the replacement of missing or dead stands and watering/irrigation when and where necessary, especially shortly after establishment. Pruning of the lowest side branches at the end of the first year of growth in the field is practised. The pruning operation should be repeated at the end of the second and third years of growth in the field. The objective of pruning is to ensure that the main stem is free of any branches up to a height of about 1 m above ground level. Branches at a height lower than this generally interfere with plantation operations.

As ruminants do not normally relish cashew stems and leaves, cattle, sheep and goats can be grazed (under supervision) in cashew plantations.

12.3.5 Control of Diseases

Cashew plants are relatively free of diseases. The mature trees are very hardy but the young seedlings are generally susceptible to the following diseases:

1. *Seedling wilt* This is a nursery disease caused by *Fusarium* spp. Seedling wilt is encouraged by overwatering. It can be controlled by reducing watering and spraying with fungicides, e.g. Captan.

2. *Leaf scab* Leaf scab occurs both in the nursery and in the field. A number of fungal species have been suspected as the causal agents. The major symptoms

Table 32 Common insect pests on cashew in West Africa

Common name	Scientific name	Family	Pest status	Damage done	Control measures
Longhorn beetle	*Analeptis trifasciata*	Cerambycidae	Major	Adult and larvae feed on stem and branch	Kill adult and larvae
Fruit piercing moth	*Achea lienardi*	Noctuidae	Minor	Feed on ripening fruit	—
Red banded thrips	*Selenothrips rubricivctus*	Thripidae	Major	Feed on leaves and young seedlings	Rogor
Flower beetles	*Pachinoda*	Scarabalidae	Minor	Feed on leaves and flowers	—
Leaf rollers	*Euproctis* sp.	Pyralidae	Major	Feed on leaves	Rogor
Leaf rollers	*Sylepta* sp.	Pyralidae	Minor	Feed on leaves	Rogor
Variegated locust	*Zonocerus variegatus*	Acrididae	Minor	Eat leaves and shoots	Gammalin 20, Unden 20
White flies	*Mesohomotoma tesmanni*	Psyllidae	Minor	Suck leaf solute	Rogor, Gammalin 20

are the development of brown and hard scabs on the under surface of the leaves. Leaf scab can be controlled by pruning to open up the canopy followed by spraying with copper fungicides.

3. *Immature fruit drop* This is common when cashew plants are grown on poor soils, especially soils which are low in potassium. It is regarded as more of a physiological rather than a pathological condition. Excessive immature fruit drop can be minimized by the application of balanced fertilizers.

4. *Fruit rot* This is a condition in which the fruits rot before ripening. The condition has been associated with a number of fungal species and it is especially common when cashew trees are cultivated in high rainfall areas. It is controlled by avoidance of high rainfall areas for the cultivation of cashew.

12.3.6 Control of Pests

Cashew plants are attacked by a number of pests. The major pests, their status, the damage done and recommended methods of control are given in Table 32.

12.3.7 Harvesting

Cashew plants under good conditions of growth come into flowering and bearing in the fourth year of field planting. Usually the first crop is small, but this increases gradually up to the tenth year when the plants come into mature production. Flowering in cashew plants is usually very profuse but only about one-tenth of the blossom sets to form fruits.

As the fruits are nearing maturity, the weed in the plantation is slashed very low to facilitate the picking of the fruits as they drop from the trees. Cashew fruits are not normally harvested, they are allowed to drop from the trees and are picked.

On gathering the fruits, the apple (see Figure 36) is cleanly removed from the seed (nut, achene). The nuts are gathered, and sun dried till they are fully dry. Drying may take three to seven or more days depending on the condition of the weather. Full drying can be ascertained by shaking the nuts together in the hands. When fully dried, there will be a sharp rattle. A dull rattle indicates that the nuts require further drying. When fully dried, nuts are sold to the processing plant.

The apple, although wasted on many plantations, can be put to a number of uses.

1. It is often sucked. The taste of the juice is slightly sharp, but pleasant.
2. The juice can be extracted and processed into a soft drink or a potable alcoholic drink by fermentation, distillation and clarification.

12.3.8 Processing

Cashew nut processing is a specialized operation. It may be done by farmers themselves or in an industrial plant. Both methods are based on the same principle, namely roasting followed by cracking and picking.

12.3.8.1 Peasant processing

There are two methods depending on the quantity of nuts to be processed.

Small numbers of nuts for individual or family use are dried and inserted in hot wood ash for up to 30 minutes; they are then cracked on a stone and the nuts gathered for use while the shell is returned to the fire as additional fuel.

When the peasant processes larger numbers for sale on the market, a native pot is used in roasting the nuts. The nuts are collected in clay pots and placed on a wood fire. As the temperature increases and the nuts begin to roast, the CNS oil in the shell of the nuts begins to 'exude' and accumulate. The CNS oil facilitates the roasting of the nuts. When the nuts are fully roasted, they are removed for cracking and picking. Fresh nuts are added to the hot CNS oil for roasting. The oil accelerates roasting of the subsequent nuts. The shells of the cracked nuts are used as fuel for roasting. Excess CNS oil can be collected and refined for industrial use.

12.3.8.2 Commercial processing

This is an industrial method based on the same principle as the peasant methods. Different cashew nut processing machines are available in the market. The essentials in a cashew processing machine are that it contains a nut cleaning chamber, grading chamber, roasting chamber, nut cracking chamber, winnowing chamber, cashew seed grading system and a bagging unit.

Commercial cashew processing is specialized and as labour is becoming more and more expensive, processing plants are being modified to meet the needs of the cashew industry.

12.3.9 Uses of Cashew Products

The cashew apple is sweet when fully ripe. The juice of the apple is rich in vitamin C and sugars. It contains 7 to 9 per cent of reducing sugars, 11 to 12 per cent of soluble solids and about 0.5 per cent of tannic acid.

The cashew nut is claimed to contain the fat-soluble vitamins A, D and K. Vitamin E occurs to a level to 200 to 210 mg/100 g. On roasting the nuts develop vitamin PP. These vitamins are known to exert a sparing action on the B group of vitamins as well as assisting in the metabolism of lactose and thiamine. The nuts also contain considerable amounts of calcium, phosphorus and iron in organic complexes. The protein content varies from 20 to 22 per cent and the quality approximates that of soya beans while distinctly ranking much superior to that of the groundnut (peanut). There is no risk of aflatoxin poisoning in cashew nut.

Finzi (1978) claimed that since cashew nut contains a reasonable quantity of linoleic acid, its consumption is a great aid in avoiding such diseases as high cholesterol levels and coronary diseases (e.g. coronary thrombosis). Cashew nut has about 80 to 82 per cent of its fats in the form of unsaturated fatty acids and as such its consumption could militate against the development of fatty liver.

The cashew apple is a good raw material for the soft drinks industry. Fermented with appropriate enzymes, it produces very valuable alcoholic drinks. Cashew can be regarded as a new crop with very great potential. More research work is needed on the culture, breeding, biochemistry, nutrition and pharmaceutical importance of both the cashew nut and the apple.

ADDITIONAL READING

Cobley, L. S. (1975). *An Introduction to the Botany of Tropical Plants.* Longman, Green & Co Ltd.

Finzi, (1978).

Komolafe, A. A. (1978). Cashew and man; *Home Echo.* Magazine of the Home Economics Department, Adeyemi College of Education, Ondo, Nigeria, pp. 86–91.

Russell, D. C. (1968). Cashew: *Extension Bulletin.* Agric. & Natural Resources, Western Nigeria.

CHAPTER 13

Oil Palm

The oil palm, *Elaeis guineensis* (Jacq.) is one of the main sources of vegetable oil. Two distinct oil types – the palm oil (mesocarp oil) and the kernel oil (the seed oil) – are obtainable from the oil palm tree. The oil palm belongs to the family Palmae. The family contains about 225 genera with over 2600 species. The oil palm belongs to the subfamily Cocoideae of which it is the most important member.

The origin of the oil palm is in dispute. Cook (1942) suggested an American origin of the oil palm, because palms grow spontaneously and freely in the coastal areas of Brazil and because the American palm (*Corozo oleifera*) has its centre of origin in Latin America. However, a number of authorities (Rees, 1965; Zeven, 19) provided fossil, historical and linguistic evidence to confirm the African origin of *Elaeis guineensis*. While some maintain a reserved position on the centre of origin of the oil palm, most authorities accept the tropical rain forest region of West Africa as the centre of origin of the oil palm.

The oil palm with various cultivars grows freely and spontaneously in this belt. Chevalier (1937) considered the area to be the 200 to 300 km wide coastal belt between Liberia and Mayumbe. Recent studies have shown that the oil palm spreads from 16 °N in Senegal to 15 °S in Angola and eastwards to Zanzibar and the Malagasy Republic. However, the main oil palm belt of West Africa runs through the southern latitudes of Sierra Leone, Liberia, Ivory Coast, Ghana, Togo, Benin, Nigeria, Cameroon and into the equatorial region of Zaire and Angola between latitudes 10 °N and 10 °S. Oil palm occurs naturally in the Semliki valley of the Zaire–Uganda border, on the Ruzizi plains between Lakes Kivu and Tanzania, on the eastern shores of Lake Tanganyika and the western shores of Lake Nyasa. Oil palm accompanied the slave trade to the New World (the Americas). It did not become widely established except in Brazil from the Ilha de Itaparia in the Bay of Salvador to the south of Marau in the State of Bahia. Oil palm plantations have been developed in a number of places in Latin America since 1960 (Purseglove, 1975). Although Malaya is one of the major producers of the oil palm, the crop was introduced to that part of the world from West Africa, especially Nigeria.

13.1 BOTANY

Elaeis guineensis grows to a height of 9 m or more, with a stout stem, covered with semi-persistent leaf bases on which epiphytes often grow. It is a monocotyledoneous and monoecious plant. The stem may be 30 to 38 cm in diameter, with progressive thickening towards the base. On older palms, the stem is punctuated with conspicuous and regularly arranged leaf scars and the stem terminates in a handsome growth of leaves (fronds). The crown may contain up to 40 or more leaves.

The palm leaf is compound and is known as the palm frond. The leaf is paripinnate with a prominent petiole (0.9 to 1.5 m long). The petiole often broadens at the base to form a clasper round the stem. Each palm frond bears from 20 to over 150 pairs of leaflets arranged in more or less two rows along each side of the flattened rachis with the longest pinnae varying up to 120 cm. The pinnae are parallel veined.

Unimproved palms flower at the age of four to five years while selected and bred palms flower in two to three years. When in full bearing, a palm may bear up to 12 bunches of fruits per year. An average mature tree may produce up to 24 fronds per year with an axillary flower cluster for each frond. The succession of production of male and female flowers is unpredictable.

The inflorescence is enclosed in a spathe, the whole structure is the spadix (see Figure 37). When fully grown, the spadix splits into two or more parts longitudinally to expose the flowers (the florets). The inflorescence is composed of a main axis and up to 40 or more side branches with florets. The male flowers are borne on side branches; they are numerous, small, each with six perianth segments arranged in two alternating whorls and an aborted ovary. The stamens consist of stout filaments each carrying a two-celled anther lobe. The anther lobes split longitudinally to release powdery pollen grains.

The female flower is composed essentially of six perianth segments which are thick and tightly folded over the pistil when young, the ovary with trifid stigmatic surfaces, and aborted stamens. Both male and female flowers are borne on thick peduncles.

There are three basic types (varieties) of the oil palm. These are as follows:

1. *The Dura type* This is characterized by thin mesocarp, thick endocarp (shell) with generally large kernels. The Dura type is genetically denoted by DD.

2. *The Tenera type* This possesses thick mesocarp, thin endocarp with reasonably sized kernel. This is a dual purpose palm for the production of mesocarp oil and kernel. It is genetically heterozygous and is denoted by Dd.

3. *The Pisifera type* This possesses thick mesocarp (with very little oil content), no endocarp (shell-less) with small kernel. The female flowers are often sterile, this resulting in bunch failure. It is genetically homozygous and recessive for shell. It is denoted as dd.

There occurred in the early days a type of palm known as the Deli palm. Deli palm was common in the Far East in the olden days. It has been assumed that it was a selection from the Tenera which was introduced into the Far East around 1840. The Deli palm has now been replaced by improved Tenera types, especially in Malaya.

The oil palm is naturally cross-pollinated. Cross-pollination is ensured by the production of two types of inflorescences – male and female – separately although occurring on the same plant; male and female flowers opening and being receptive at different times. Rarely do the terminal female flowers of an

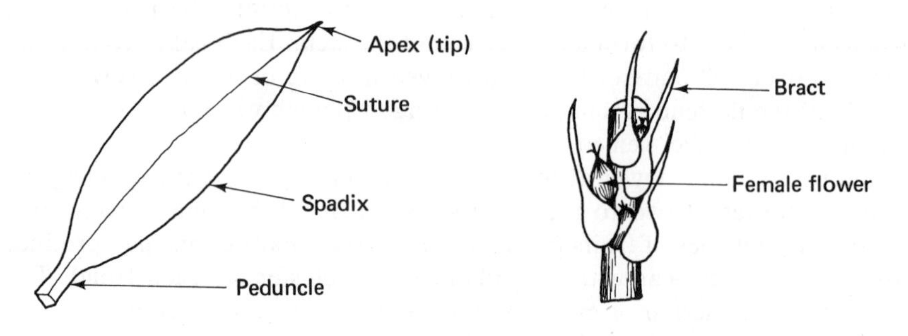

(a) The spadix

(b) A piece of the inflorescence of the oil palm

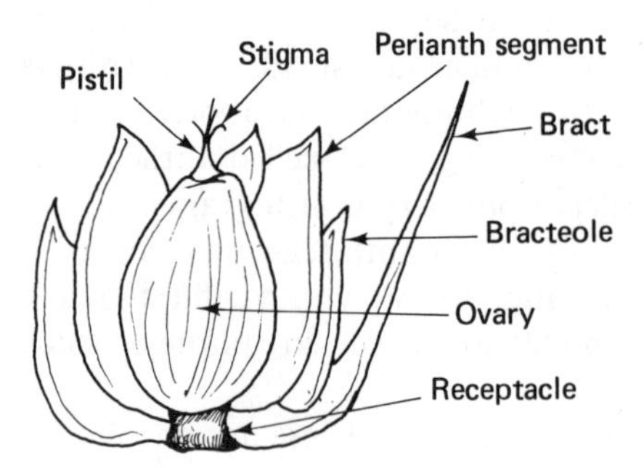

(c) Longitudinal section of female flower of the oil palm

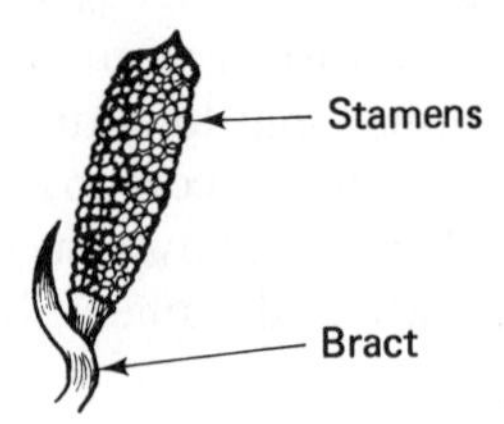

(d) Branch of male inflorescence

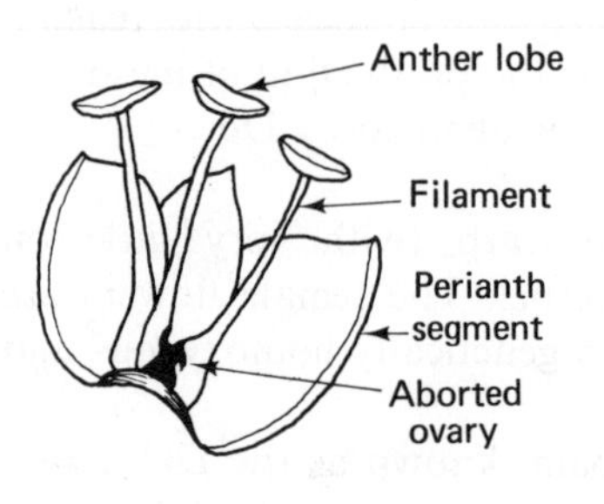

(e) Longitudinal section of a male flower of the oil palm

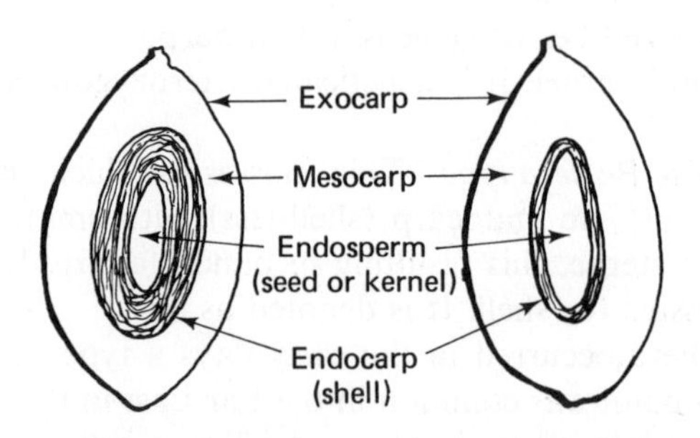

(f) Longitudinal section of a thick-shelled fruit (a varietal) type of the palm oil

(g) Longitudinal section of a thin-shelled fruit (a varietal) type of the palm oil

Figure 37. Floral structures of oil palm

inflorescence still remain receptive when the male flowers of the next upper ring of the inflorescence commence discharging pollen grains. The main agencies of pollination are wind and insects, especially those of the **Diptera**.

Immediately after fertilization, fruit development commences with rapid expansion of the husk, enlargement of the cavity followed by the development of the endosperm and the embryo. The embryo is embedded in the endosperm just below the eye. Initially the embryo is thin and jelly-like, but later thickens to fill the cavity and hardens. The inflorescence develops into the fruit bunch.

13.2 THE ENVIRONMENT OF THE OIL PALM

The oil palm belongs in the warm, high rainfall, tropical forest areas. It grows best where rainfall is not less than 1500 mm, evenly distributed throughout the year. Ideal temperatures are 27 °C to 35 °C, while the nights should be frost free. Oil palm will tolerate even higher temperatures provided there is adequate moisture.

Oil palms require plenty of sunshine, productivity being reduced in areas with excessive sky overcast. The oil palm thrives under conditions of high relative humidity; yields are adversely influenced when the crop is exposed to dry harmattan winds.

Oil palm is a lowland crop although it can grow well up to an altitude of 900 m. It has a fibrous root system and benefits from deep soils which are fertile, free from iron concretions and well drained. It could also tolerate a fair range of soil pH although neutral soils are most favoured.

13.3 AGRONOMY OF THE OIL PALM

13.3.1 Selection of a Site

The use of any piece of land for planting oil palm is determined by the quality of the soil and the topography. Land for planting oil palm should be level or slightly undulating, well drained, fertile and deep. Virgin forest soils are preferable to soils that have been cropped. When land that has been previously cropped is used for growing oil palm, appropriate fertilizer must be provided.

13.3.2 Propagation

Oil palm is propagated by seed. Seeds being currently issued to farmers for planting are derived from hand pollination of Dura × Pisifera palms which give 100 per cent Tenera type thin-shelled, dual purpose palms. Seeds are issued in two grades – the grading depending on the fruit and bunch composition of the original parents. In Nigeria, a minimum average bunch yield of 68 kg per palm per annum is regarded as the standard for all grades. The two grades of seeds are special grade and ordinary grade.

The specifications for the special grade are as follows:

The Dura parent should have the following minimum composition:

1. Fruit to bunch ratio 65 per cent.
2. Shell to fruit ratio 35 per cent.
3. Mesocarp and kernel to bunch ratio 45 per cent.

The Pisifera parent (the pollen parent) must have been progeny tested in crosses with good quality Dura palms.

The ordinary grade of seed is obtained from Dura parents, selected from two groups. The first group may be taken from unknown parentage with the following minimum standards for bunch quality:

1. Fruit to bunch ratio 63 per cent.
2. Shell to fruit ratio 28 per cent.
3. Mesocarp and kernel to bunch ratio 42 per cent.

The second group consists of the best palms from progenies with the following minimum bunch composition:

1. Fruit to bunch ratio 60 per cent.
2. Shell to fruit ratio 38 per cent.
3. Mesocarps and kernel to bunch ratio 40 per cent.

13.3.2.1 Collection and preparation of seeds

Under normal conditions, oil palm seeds are de-pulped by subjecting the fruits to heat followed by de-pulping. Seeds intended for planting should not be subjected to high temperatures for de-pulping purposes. The mature bunches are harvested, the seeds are extracted from the bunch, collected in a cool place till the mesocarp softens on its own, then the kernels are washed clean in cold water, air dried and stored or germinated as outlined below.

13.3.2.2 Germination of seeds (pre-nursery)

Oil palm seeds germinate so slowly in nature that it has become necessary to develop special methods to enhance the germination process. The basic principle in oil palm seed germination is to subject the seeds to high temperatures under controlled conditions to induce germination. Two methods are used. These are the wet heat method, which is no longer popular, and the dry heat method, which is currently very popular. This latter method consists of subjecting the seeds to heat treatment at moisture contents too low for germination to take place, germination occurring later when the moisture is restored to optimal level. The process takes longer than the wet method, but it ensures more uniform germination of the seeds. In outline, the procedure is as follows:

1. Seeds at 15 per cent moisture content are placed in intact 500 gauge polythene bags at the rate of 750 seeds per bag of 60 × 60 × 120 cm.
2. The bags are secured with rubber bands, allowing a good air space on top of the seeds (see Figure 39a) page 266.
3. The polybags are arranged in wooden boxes.
4. The boxes are then placed in a germinator maintained at 39 °C to 40 °C for 75 to 80 days, or at other temperatures or periods as recommended by the seed supplier.
5. After the heat treatment, the seeds are soaked in cold water for three days, the water being changed every 24 hours. After this treatment, the moisture content of the seeds will be 28 to 30 per cent for Tenera, or 21 to 22 per cent for Dura.
6. The seeds are then drained, and dried under shade until the water film adhering to each seed disappears (one to two hours depending on the ambient temperature and relative humidity).
7. After complete draining, the seeds are returned to the polybags and kept in a cool place at room temperature.
8. Seeds are then examined every two weeks for drying and germination. If drying is noticed, they are lightly sprayed with water, while germinated seeds are picked out for potting. Vigorous germination normally occurs about 15 to 21 days after the post-heating soaking. This method gives about 80 per cent seed germination.

It is usual to reheat seeds that do not germinate after 21 days for another 20 days and repeat the process to get them to germinate. Although germinated seeds could be planted out into the field, it is not advisable. Germinated seeds should pass through the nursery.

13.3.2.3 Nursery

The type of care that seedlings require during the early period of growth is better provided in a nursery.

A popular method is the single stage polybag nursery. This system involves the use of 400 to 500 gauge black polythene bags measuring 40 cm wide × 35 cm deep, which are filled with topsoil. The bags are placed at 45 cm square spacing. The germinated seeds are planted into these polybags and thickly mulched with shredded partially decomposed oil palm bunch refuse. Each polybag is perforated at the bottom to allow excess water to drain out.

Polybag nursery seedlings are easily maintained by watering with 1.5 litres or water applied by hand per week, especially during the dry season. A solid mixture of NPK Mg fertilizer in the ratio of 1 : 1 : 1 : 1 using sulphate of ammonia, muriate of potash, single superphosphate and magnesium sulphate should be applied twice at the ratio of 56 g per seedling when the seedlings are two and eight months respectively. Dithane M45 is used to spray against diseases

Figure 38. Polybag palm ready for transplanting (Courtesy of the Nigerian Institute for Oil Palm Research)

at fortnightly intervals. The polybag nurseries should be clean-weeded to eliminate competition for water and nutrients, and to reduce tension on the bag caused by the root growth of the weeds. Shading is unnecessary. Seedlings are transplanted into the field in April/May or when the early rains are regular (see Figure 38).

Polybag nursery seedlings are usually ready for transplanting into the field when they are 10 to 12 months old.

13.3.3 Land Preparation

The standard method of land preparation for planting oil palm is clear-felling. The trees and trash are burnt. The field is blocked, and each block is lined out and the planting holes dug.

In felling trees, efforts should be made to make them fall along the same direction, e.g. north-south direction; this makes movements of machinery on the farm easy. The standard spacing for oil palm is 8.7 m triangular. This permits early intercropping of the land before the oil palm develops its canopy.

13.3.4 Transplanting of Oil Palm Seedlings

Planting into the field usually starts at the onset of the rains. Seedlings planted late stand very poor chances of survival in the field since they may not establish themselves before the onset of the dry season, so the earlier field planting is completed, the better.

Polybag seedlings are best transplanted with the entire soil in the polybag. During planting, care is taken to remove the polythene bag without injuring the mass of ramifying roots. The seedling is planted into a hole slightly wider than the size of the polybag by burying the entire ball of earth obtained after the removal of the bag so that the surface of the ball is level with the surrounding soil. It is very important to see that the base of the seedling is above the surrounding soil; deep planting leads to failure of the seedling to develop properly or may even lead to death.

Where plantations are too distant from the nursery and transportation costs are likely to be prohibitive, polybag seedlings may be transplanted with naked roots. After removal of the bag, the roots are carefully washed free of soil and dipped in clay slurry. The seedlings may be tied together in bundles of 10 with their roots well protected with wet grass, palm leaves or banana fibre or they could be placed in a deep basket or other container lined with wet grass or wet sacking. Seedlings raised in field nurseries are also transplanted with naked roots.

In planting naked-roots seedlings, the planting holes are shaped like a cross with a slight elevation at the centre (see Figure 39). The roots are evenly distributed in the four arms of the cross, and the hole filled with topsoil and firmly consolidated.

After planting the young palms should be protected from damage by animals, such as rodents and grass cutters, that love eating the heart of the young palms. This could be done by planting a collar of wire netting around each seedling and pegging it down. The netting should be cut 45 cm high by 12 cm long. This collar will sit about 15 cm away from the base of the palm, and encircle it completely.

13.3.5 Post-planting Maintenance Operations

The establishment of a leguminous cover crop is desirable. This follows the same principles as for other crops.

Other routine matters to be arranged are weeding, supply of vacant or dead stands, mulching and removal of undecomposed mulch during the early rains. Watering/irrigation of newly planted seedlings must be carried out as necessary. The removal of the lower senescent leaves is beneficial. Such leaves should be removed, gathered outside the plantation and burnt.

13.3.6 Nutrient Deficiencies and Fertilizers

Although oil palm grows on soils of comparatively low fertility, it is particularly susceptible to deficiencies in nitrogen, potassium and magnesium. When any of these elements is deficient, the yield of fruits is seriously reduced, and fruiting may cease. Magnesium deficiency may be aggravated or induced by the application of potassium fertilizers. Nitrogen is likely to be deficient during the first three years in the field.

Consistent economic benefits have been obtained from applications of potassium, magnesium and nitrogen fertilizers. Responses have sometimes been obtained from application of phosphate fertilizers. Nitrogen is essential for the early vegetative growth of newly established oil palms. Investigation into trace element nutrition has shown the incidence of boron deficiency to be fairly common while copper deficiency may occur in peat soils.

13.3.6.1 Nutrient deficiencies

13.3.6.1a Potassium deficiency

Potassium deficiency is usually marked by a drop in yield. The first visual symptom is the appearance of minute, pale green rectangular spots on the leaflets of older leaves, which expand as the affected leaves become older. As the spots expand into round or rectangular shapes, their colour changes through olive-green to a bright orange, and as adjacent lesions fuse they form the compound orange spots, hence the name of the symptom 'confluence orange spotting'. In severe cases terminal and marginal necrosis of affected leaflets results, and the tissues of the leaflet become pale greyish brown in colour, dry, brittle and later disintegrate leaving the leaflet with a ragged irregular margin. Severe potassium deficiency can be corrected by an initial application of 2.25 kg of potash per palm, followed by annual dressings of 0.66 kg per palm.

13.3.6.1b Magnesium deficiency

The most characteristic symptom of magnesium deficiency is a uniform discoloration from yellow to orange-yellow, appearing 10 to 12.5 cm behind the tip of the leaflets of the oldest leaves – usually those parts most exposed to light. A characteristic feature of magnesium deficiency is the absence of chlorosis in portions of leaflet tissue protected from direct sunlight by shading. The upper rank leaflets first show the yellowing followed later by leaflets in the lower rank.The leaflets may become diseased, the tips of the leaflets turn purplish and dark brown and finally dry out from the leaf tip inwards. The deficiency is characteristically known as 'orange frond'. Magnesium deficiency can be controlled by an initial dressing of about 2.25 kg of Epsom salt (magnesium sulphate) per palm, followed by 0.66 kg per tree per annum.

13.3.6.1c Nitrogen deficiency

Symptoms of nitrogen deficiency are more common in young palms. The affected fronds first become pale green in colour and later pale or bright yellow as the chlorosis becomes more severe. The tissues of the midrib become bright yellow or orange in marked contrast to the paler chlorosis of lamina tissue. Both the upper the lower rank leaflets are equally affected by the chlorosis.

13.3.6.2 Fertilizer regimes

Young palms benefit immensely from the application of sulphate of ammonia at the rates shown in Table 33.

If the need for nitrogen fertilizer has been established for older palms, 1 kg of sulphate of ammonia should be applied each year to each palm at the beginning of the rainy season. The fertilizer should be broadcast on a circular clearing 2 m in diameter around the palm.

1. *Potassium* Applications of potassium fertilizers should continue throughout the life of the palm in the field. Normal dressings of potassium chloride (muriate of potash) or potassium sulphate (sulphate of potash) should be as shown in Table 34.

In subsequent years 0.66 kg potassium chloride or potassium sulphate could be applied to each palm annually, at the beginning of the rainy season.

In some areas where the soil has been seriously leached, regular application of both sulphate of ammonia and sulphate of potash may cause symptoms of magnesium deficiency to appear, in which case a balanced quantity of magnesium sulphate should also be applied.

Table 33 Rate and age of application of ammonium sulphate to palms

Age of palm	Time of application	Rate of application
Year of planting	(a) 6 weeks after planting	0.25 kg/palm
	(b) 6 months after planting (in October)	0.25 kg/palm
1st year after planting	Early in rains	0.5 kg/palm
2nd year after planting	Early in rains	1 kg/palm

Table 34 Rate and age of application of potassium fertilizer to palms

Age of palm	Time of application	Rate
Year of planting	6 weeks after planting	0.25 kg/palm
1st year after planting	Early in the rains	0.5 kg/palm
2nd year after planting	Early in the rains	0.66 kg/palm

Table 35 Rate and age of application of ammonium sulphate, potassium sulphate and magnesium sulphate to palms

	Year												
	Establishment years							Yielding years					
	Year of planting		1	2	3	4	5	6	7	8	9	10	11
Fertilizer	6 weeks after planting	6 months after planting	May	May	May	May	May	May	May	May	May	May	May
	kg	kg	kg	kg	kg	kg	kg	kg	kg	kg	kg	kg	kg
Ammonium sulphate	0.25	0.25	0.25	1									
Potassium sulphate	0.25		0.5	0.66	0.66	0.66	0.66	0.66	0.66	0.66	0.66	0.66	0.66
Magnesium sulphate (Epsom salt)	0.25		0.5	0.66	0.66	0.66	0.66	0.66	0.66	0.66	0.66	0.66	0.66

2. *Magnesium* When the magnesium status of the soil is low, it could be due to: (*a*) absolute deficiency of available magnesium; or (*b*) antagonistic effects of other nutrient elements like potassium or nitrogen, inducing magnesium deficiency. The situation can be corrected by applying 0.25 kg, 0.5 kg and 0.66 kg respectively of magnesium sulphate per palm in the first three years after planting followed by 0.66 kg per palm per year in subsequent years.

3. *Phosphorus* There is no general recommendation for phosphate manuring of oil palm, it is only in one area in Ghana that striking responses have been obtained.

A typical scheme of fertilization for oil palm applicable to most areas of West Africa is shown in Table 35. The rates are in quantity per palm per application. In drier areas, sulphate of ammonia may need to be applied at the rate of 1 kg per palm up to the sixth year.

13.3.7 Rehabilitation of Oil Palm Plantations

Rehabilitation of oil palm plantations may be necessitated by a number of factors such as diseases, pests, old age of trees or the need to change to a new and more productive variety.

In the case of disease- and pest-infested farms, the old trees are uprooted and removed from the plantation or burnt. The remains (roots, broken fronds and other debris) are also gathered and burnt. A cover crop is established and the land rested for at least one year as an additional means of eradicating the offending disease or insect pest. Subsequently, planting holes are dug at points intermediate to old plant stands and new seedlings are planted.

In rehabilitating a disease- and pest-free old farm, the standard practice is to plant young seedlings under the old ones. As the young trees begin to develop, the old ones are removed. This is also the procedure when rehabilitation is necessitated by the need to change to a new variety.

When part of the trees is damaged, rehabilitation consists of uprooting the damaged tree (trees), planting a new seedling in its place, applying fertilizer, mulching, and ensuring that insect pests do not invade the palms around the gap created in the canopy.

13.3.8 Diseases of the Oil Palm

The most common diseases of the oil palm are listed below.

1. *The brown germ, caused by Aspergillus spp.* This is very common and widely distributed throughout the oil palm growing areas of the world. The main symptoms are brown spots appearing on emerging buttons, spreading, coalescing and the affected tissues becoming shiny and rotten. For controlling the disease the dry heat method of germinating seeds should be used, as wet conditions encourage fungal growth. Seeds should be treated with fungicides before germination.

2. *Anthracnose* This is a general term referring to a complex disease condition in the oil palm. Anthracnose is believed to be caused by various fungi belonging to the following genera: *Botryodiplodia, Melanconium, Glomerella* and *Corticium.* The general condition of the disease is the appearance of dark necrotic lesions on the leaves of seedlings usually at the pre-nursery and nursery stages. The appearance of the necrotic lesions depends on the specific causal organism

(a) *Botryodiplodia palmarum* is characterized by small translucent spots, typically near the tip or edge of the leaf or where the leaf is damaged by insects or other pests. At an advanced stage, the colour changes to dark brown, surrounded by a yellow halo or a transitional zone. On drying out the lesions enlarge and the centres turn grey. The pycnidiophores are responsible for spreading the disease.
(b) *Melanconium elaeidis* symptoms are similar to those of *B. palmarum*, but the lesions are lighter brown with a pale yellow halo and the lesions dry out more rapidly. Acervuli develop with spherical spores to spread infection further.
(c) *Glomerella cingulata* is characterized by long lesions between the veins. The lesions do not normally cross the veins. Brown and black necrotic tissues initially appear water-soaked and are surrounded with a yellow halo. Acervuli with conidia appear on the dry lesions. The acervuli contain flask-shaped perithecia, from which ascospores are released.
(d) *Corticium solani* has also been accused of causing anthracnose. Anthracnose is a disease to be watched for when palm seedling are newly transplanted into the field. The causal fungi are all weakly parasitic, and therefore their outbreak can be prevented by sound agronomic practices and adequate farm sanitation. When there is an outbreak, fungicides like Captan, Ziram or others should be applied.

13.3.9 Pests of the Oil Palm

The oil palm, like the coconut palm, has a lower number of pests than most tropical tree crops such as cacao and kola. The major pests of the oil palm, the damage done and possible methods of control are listed in Table 36.

13.3.10 Harvesting

A palm bunch is ready to be harvested when it has just a few loose fruits. It is essential that each palm is inspected at least once a fortnight for ripe bunches as over-ripe fruits produce lower quality palm oil.

The major factors which affect the yield of the oil palm include the successful production of a large number of female spadices, the number of fruits set per

Table 36 Pests of the oil palm in West Africa

Name of pest	Group	Damage done	Method of control
Mites	Acaridae	Damage during germination especially if charcoal boxes are used. Germ spores of seeds are destroyed.	Spray with Rogor. Maintain high level of sanitation
Red spiders	Acaridae	Attack leaves; leaves turn bronze in colour, die prematurely. Attack favoured by dry weather	Treat plants with Tedion-V-8 Wettable powder 20% or emulsion 8%, or Rogor 40.
Grasshoppers, locusts	Acridoidea	Attack leaves and growing points by feeding on them	Spray with appropriate insecticide
Termites	Isoptera	Attack usually on unhealthy plants. Damage roots and senescent leaves	Treat with Rogor 40
Aphids	Hemiptera	Feed on young shoots. Attack mainly in the nursery	
Mealybugs, scale insects	Coccidae	Attack generally in the nursery. Difficult to control by spraying	Use systemic insecticide or heat scarification
Spear borer	Pyralidae	The African spear borer attacks young leaves and growing points both in the field and in the nursery	Spray with DDT or Rogor 40
Rodents, monkeys, flying bats	Mammals	Feed on ripe fruits	Harvest ripe fruits early. Shooting and trapping
Weaver birds	*Quelea quelea*	Could be very serious. They nest in swarms on the palms completely defoliating them	1. Aerial spray with avicides 2. Shooting to scare them away

bunch, the size of each fruit which is genetically determined although profoundly affected by the environment, the mesocarp/kernel ratio, the oil content of both the mesocarp and the kernel, and finally the fruit/bunch ratio.

There are three methods of harvesting palm bunches. In all cases, it is essential that damage to leaves be restricted. Only those leaves that hinder removal of the bunch should be cut.

1. *Chisel method* The chisel consists of a piece of flat iron 23 cm long (an old lorry spring will do) one end of which is rounded off and well sharpened. The other end is bolted to one end of a metal water-pipe 23 cm long. Inside the hole at the other end of the water pipe is fixed a wooden handle up to 0.75 m

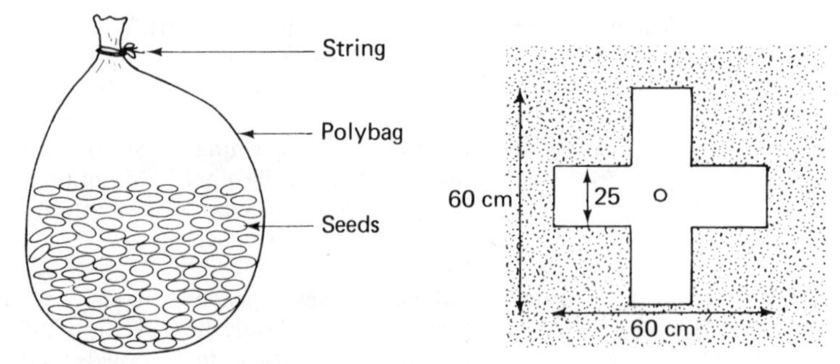

(a) Polybag containing seeds preparatory to heat treatment

(b) Planting hole for naked-root palm seedling

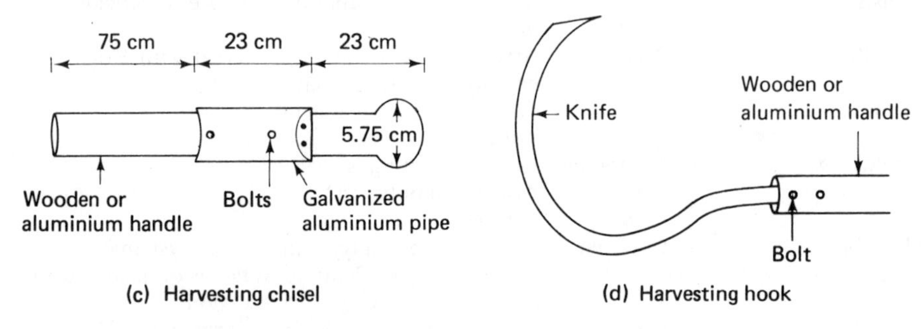

(c) Harvesting chisel

(d) Harvesting hook

Figure 39. Germination and harvesting. Aspects of the oil palm

or more long. The implement can be made by any village blacksmith (see Figure 39). The harvesting chisel is used for harvesting bunches from the time the palms first come into bearing until the palms become too tall for the use of this implement.

When a bunch is to be harvested the wooden portion of the chisel is held with two hands and the sharpened end of it is used to strike hard at the stalk of the bunch. If the stalk is not sufficiently cut by this one strike it is repeated and the bunch then pushed out of the palm. A good harvester needs only one strike and by careful manipulation of the implement can have the stalk cut and the bunch pushed out. In removing the bunch there should be no need to cut any leaves.

To avoid harming the stem of young palms just coming into bearing care is required in the use of the harvesting chisel.

2. *The pole-knife method* The pole-knife, often referred to as the harvesting hook, is used in harvesting bunches from palms which have become too tall to be harvested with the chisel. A Malayan knife, sickle-shaped, is firmly tied on to a pole (Indian bamboo or any strong 'bush' pole) with binding wire. The length of the pole depends on the height of the palms to be harvested (see Figure 39). The knife is usually well sharpened, and a sheath should be provided to cover the knife when the implement is carried by a harvester along any roads.

When a bunch is to be harvested with the harvesting hook the harvester stands at a convenient spot to enable him to get at the stalk of the bunch. If the bunch to be harvested is subtended by one leaf or more which prevent access to the stalk, the leaf is cut off close to the tree trunk with the knife. Then the harvester 'hooks' the stalk with the knife and gives a good pull downwards. A good harvester usually succeeds in getting the bunch down with one such pull. At times he may, after the cut, hook the top of the bunch and then pull downwards so that the bunch can fall to the ground. A well-trained man can harvest palms of up to 8 m high or more using the pole-knife method. As all trees in a plantation cannot be of equal height it is advisable for a harvester to carry two sets of hooks, one on a short and one on a long pole so that whichever is suitable can be used when harvesting.

3. *Climbing ropes* The method of harvesting palm bunches by climbing with ropes is well known. The system is dangerous, slow and expensive. It is advisable that before this stage is reached, all the tall palms should be felled and the area replanted.

13.4 PROCESSING OF PALM FRUITS

The operations involved in the processing of palm fruits to palm oil and kernels can be briefly summarized as follows:

Sterilization → stripping → milling → separation →
pressing → clarification → storage or sale of palm oil

After separation, the kernels may be subjected to drying, cracking, grading and bagging for storage or sale, if they are not to be processed immediately.

13.4.1 Extraction of Palm Oil from the Mesocarp

13.4.1.1 Sterilization

Sterilization of palm fruits or bunches before milling for the purpose of oil extraction, is achieved by boiling. Sterilization softens the fruits for stripping from the bunches and for milling; it disinfects the fruits by killing pathogens, and inhibits the action of lipolytic enzymes. Sterilization can be carried out in pots, drums or in sterilization chambers as is done in the oil mills.

13.4.1.2 Stripping

This is the removal of fruits from sterilized or quartered bunches. The stripped fruits are re-sterilized for about 30 to 45 minutes before milling (pounding). Fruits pound more easily when they are hot and well sterilized, i.e. softened.

13.4.1.3 Milling

This is the pounding of sterilized fruits for the purpose of separating the mesocarp from the kernel (de-pulping). After the separation, the mesocarp is pounded until no streak of coloured outer skin is distinguishable any more.

13.4.1.4 Pressing

The pounded mass is then loaded into a press for the extraction of the oil. At times, it may be necessary to add water to the mass for softening purposes to facilitate oil extraction. There are different types of presses: the screw hand press, hydraulic press and centrifugal press.

13.4.1.5 Clarification

In the traditional method of oil palm fruit processing, the extracted crude oil is clarified by boiling and skimming. With the use of modern hydraulic and hand presses, the extracted crude oil is too heavily laden with sludge to be properly clarified in the traditional method. Specially constructed double jacketed clarification drums are used.

Two clarification drums (each double jacketed) are required for efficient clarification. The process is as follows:

1. The drums are mounted over an open fire.
2. About 45 litres (10 gallons) of water is poured into each of the outer drums and brought to boil.
3. The crude oil is introduced. The crude oil will flow through the boiling water and deposit the sludge while the oil floats on top of the water. As more crude oil is added, the refined oil will overflow into the inner drum. It is important that the contents of the drums are kept as hot as possible without causing boiling throughout the operation. Boiling stirs up the sludge from the bottom of the drum and may lead to the formation of an unwanted emulsion of oil and water. Clean oil is withdrawn from the inner drum, usually termed No. 1. The sludge is fed to the outer drum and the process is repeated till the oil is completely clarified from the sludge.

At the end of a day's operation, hot water should be used to bring up the level of oil in both drums until the oil is completely swept off. The oil obtained is then simmered or fried over low fire to remove traces of water. The oil thus refined is stored in drums, tankers, tins or bottles, ready for sale.

13.4.2 The Palm Kernel

After separation from the mesocarp, the kernels are washed, dried in the sun, cracked by hand and packed for sale.

Under mechanical processing, the kernels on separation are conveyed into a drying chamber where they are rapidly dried. Next they are passed on to the crackers. On cracking both the shells and kernels are partially separated. Final seperation is achieved by passing the mass into a clay slurry where the kernels float and the shells sink. The kernels are collected, cleaned, dried and packed for sale.

The quality of the palm kernel is determined by the following:

1. The free fatty acids (FFA) content, the less FFA the higher the grade.
2. The colour of the kernel, kernels which are light yellow in colour are preferred to dark-coloured ones.

Palm kernels are, therefore, traded on their FFA content; the amount and extent of external and internal discoloration; the moisture content (6 per cent or less is recommended for storage); the presence or absence of mould and other foreign contaminants; and the number of broken kernels (the less the higher the grade).

13.4.3 Quality of Oil Extracted from Oil Palm Fruits

The method of extraction determines the type of oil produced. Palm oil is classified roughly as a soft oil (with low FFA content), hard oil (with high FFA content) and special oils (very low FFA content).

13.4.3.1 Hard oil

The bulk of oil produced in the traditional oil producing areas of the world is hard oil. In this process, the ripe fruits are harvested, the bunches are heaped or cut into sections before heaping. The heap is left until the individual fruits are loose and fall out of the bunch. The fruits are picked. At this stage, one of two processes can be adopted:

1. The fruits can be heaped in a dug-out or a pit and allowed to rot. Then they are mashed by treading and water added to skim off the oil. The skimmed oil is then boiled and clarified. This process usually results in oil with high FFA content.
2. The fruits can be boiled for softening, then pounded and the mass separated from the kernel in water, the mesocarp fibre washed out, the oil skimmed off and clarified by boiling. This process produces oil with somewhat lower FFA content.

13.4.3.2 Soft oil

The ripe bunches are harvested, cut into sections (quartered), heaped and covered with leaves for not more than five days with daily sprinkling of water especially during the dry season. The fruits loosen rapidly and are collected, boiled, pounded, separated by immersing the pounded mass in water; the crude oil is skimmed off and clarified. This is followed by frying to remove traces of water. This process usually produces oil with very low FFA content.

13.4.3.3 Special oil

In modern oil palm plantations, fruits are mechanically processed immediately on harvesting in specially constructed and powered mills such as the Poineer or Stork Oil Mills. In operating the mills, the fruit bunches are fed to the mills and all processes are automatically carried out until the special oil flows out and the kernel and shells are discharged through a separate outlet. Some mills are coupled with kernel driers and crackers.

13.4.4 Efficiency of Oil extraction

Efficiency of oil extraction varies with different methods. It is lowest in the native methods of processing hard, oil, intermediate with the hand and hydraulic presses while it is very high with the mechanical mills. Efficiency of the various operations is calculated as follows:

1. Estimated oil extraction efficiency

$$= \frac{\text{Wt. of oil produced} \times 100}{\text{Wt. of oil produced} + \text{losses accounted for}}$$

2. Estimated kernel extraction efficiency

$$= \frac{\text{Wt. of kernel produced} \times 100}{\text{Wt. of kernel produced} + \text{losses accounted for}}$$

3. Efficiency of kernel cracking

$$= \frac{\text{Dry wt. of kernel produced} \times 100}{\text{Dry wt. of nuts entering nut drier}}$$

Losses of palm oil are mainly due to loss in mesocarp refuse, to remains of fruits in bunch refuse and to inefficiency of mills, e.g. leakages, and defective milling.

Losses of kernels are mainly due to loss in bunch refuse, or mesocarp fibre, kernels sticking to the shells and to uncracked nuts remaining among the shells.

13.4.5 Composition of Palm Oil and Palm Kernel oil

13.4.5.1 Palm oil

Palm oil is composed mainly of palmitic acid and oleic acid, hence saturated and unsaturated fatty acids roughly in equal proportions (see Table 37). In

Table 37 Average composition of constituent fatty acids in palm oil

	Per cent
Stearic acid (C_{18})	4.30
Palmitic acid (C_{16})	41.20
Myristic acid (C_{14})	2.30
Linoleic acid	9.60
Oleic acid	42.50
Palmitoleic acid	1.10

addition, there are traces of lauric and linolenic acids. Apart from the fatty acids, the major constituents of palm oil are the carotenoids – mainly carotene which gives the palm oil its red coloration.

The quality of palm oil is determined by the following characteristics:

1. The content of free fatty acids (FFA). Free fatty acids arise due to autocatalytic action of fats in oil, by lipolytic enzymes from bruised or damaged fruits or by microbial lipases present as contaminants. Most of the FFA are formed in fruits prior to processing due to damage to the fruits.
2. Absence of contamination with water or other impurities. Presence of water and other foreign contaminants encourage high rates of FFA formation through the activity of the lipolytic enzymes (lipases).
3. High bleachability – this is very important in oil destined for industrial purposes.

13.4.5.2 Palm kernel oil

Two main products are obtainable from the palm kernel in addition to the shells. These are the palm kernel oil and the palm kernel cake, which is extensively used in livestock feed.

The palm kernel oil is similar to coconut oil with which it is interchangeable.

Table 38 Average fatty acid composition of palm kernel oil

	Per cent
Caprylic acid (C_8)	3.00
Capric acid (C_{10})	6.00
Lauric acid (C_{12})	50.00
Myristic acid (C_{14})	16.00
Palmitic acid (C_{16})	6.50
Stearic acid (C_{18})	1.00
Oleic acid	16.50
Linoleic acid	1.00

It contains a high proportion of saturated fatty acids with small amounts of the lower molecular weight acids (see Table 38).

Palm kernel cake (PK meal) is obtained after milling the palm kernels. Its composition varies with the efficiency of its oil extraction. An average composition is recorded in Table 39.

13.4.6 Essential Precautions in Processing Palm Fruits

The aim in processing palm fruit is to obtain high quality oil which attracts the highest price in the market. This aim can be achieved by observing the following precautions when processing fruits:

1. Fruits should not be wounded during harvesting, transportation to the mill and during feeding of bunches to the mill.
2. Fruits should be harvested when they are just ripe. Over-ripening should be avoided.
3. The barest minimum time interval should be allowed between harvesting and processing (sterilization).
4. The processing system must ensure that the material remains at high temperature throughout the period of processing.
5. Processed oil must be stored in clean, dry and sterilized containers.

Table 39 Average composition of palm kernel cake (PK meal)

	Per cent
Carbohydrates	48.00
Oil	5.00
Proteins	19.00
Fibre	13.00
Ash	4.00
Water	11.00

Table 40 Percentage FFA contents (palmitic acid) in oil after various periods at three temperature levels

Time/temperature	0.25 per cent water in oil	5.25 per cent water in oil
At extraction	2.80	2.80
After 62 days at 18 °C	3.30	3.30
After 55 days at 55 °C	3.50	4.10
After 36 days at 75 °C	4.60	9.70

(See C. W. S. Hartley, 1970, *The Oil Palm*)

Oil in storage usually deteriorates through autocatalytic hydrolysis. The traces of fatty acids present act as catalysts in the reaction between the triglycerides and water. In practice autocatalytic hydrolysis stops almost completely when the moisture content of processed oil is kept below 0.10 per cent. Table 40 shows the relations of autocatalytic hydrolysis with water contents and temperature ranges in palm oil.

Contamination of processed oil with foreign matter and microorganisms, is known to result in oil spoilage through lipolytic activities of microorganisms. Therefore, it is absolutely essential to maintain a very high level of sanitation during the processing and storage of palm oil.

13.4.7 Quality Control in Oil Palm Mills

High quality is indispensable in all products derivable from the palm fruit. In order to ensure such a high quality, it is standard practice to attach a quality control laboratory to palm oil mills. The main functions of the laboratory are to analyse the major products for their quality characteristics.

1. *Palm oil* The palm oil is analysed for its moisture content in standard laboratory manner, FFA content (by titration against sodium hydroxide) and iodine value.

2. *Palm kernel* The palm kernel is analysed for amount of pieces of shell, broken kernels, whole kernels, moisture and FFA content.

13.4.8 By-products

By-products of the oil palm include the following:

1. Bunch refuse which is used as fuel especially in the Pioneer Oil Mills. The residual ash is high in potassium and is used as fertilizer.
2. Central axis of the inflorescence which can be used as fuel, but more often it is retted, washed and used as a sponge or in making ornamentals. Both the bunch refuse and the floral axis can be used for mulching or for soil conservation in filling trenches or as silt catchers.
3. The palm frond is used for building, especially as roofing material and fencing, the midrib of the leaflets is used for brooms and toothpicks. The frond when dry is an excellent fuel (firewood).

13.5 PALM WINE

This is the delicious wine obtainable by tapping the base of the immature inflorescence of the oil palm. It is very similar to toddy in chemical composition.

Freshly tapped, undiluted and chilled, palm wine is a very pleasant drink during the cool hours of the day. It is also very high in yeast content.

ADDITIONAL READING

Adansi, M. A. (1977). Prospection and exploitation of genetic resources of oil palm as a basis for improvement. *Proc. AAASA Symposium on Plant Genetic Resources.*

Annual Report of the Nigerian Institute for oil Palm Research 1964–1976.

Ataga, D. O. (1973). Determining moisture content of some soils supporting the palm (*Elaeis quineensis*) by the neutron modern method. *Journal of NIFOR*, **5** (18), 13–21.

Aya, F. O. (1974). The use of polythene bags for raising oil palm seedlings in Nigeria. *Journal of NIFOR*, **5** (19), 7–12.

Chevalier, A. (1937). On nouveau cafeier sauvage de Madagascar a graine sans cafeine. *Rev. Bot. Appl.*, **195**, 821–826.

Hartley, C. W. S. (1970). *The Oil Palm*; Longman, London.

Obasola, C. O. (1973). Female sterility and fertility in variety Pisifera and shell thickness in progenies of Dura × Pisifera crosses of the oil palm (*Elaeis guineensis* Jacq.). *Journal of NIFOR*, **5** (18), 37–41.

Obasola, C. O. (1973). Breeding for short-stemmed oil palm in Nigeria. (1) Pollination, compatibility, varietal segregation, bunch quality, and yield of F_1 hybrids *Corozo oleifera* × *Elaeis quineensis*. *Journal of NIFOR*, **5** (18), 43–53.

Otedoh, M. O. (1974). Raphia oil: Its extraction, properties and utilization. *Journal of NIFOR*, **5** (19), 45–49.

Purseglove, J. W. (1975). *Underexploited Tropical Plants with Promising Economic Value*. National Academy of Sciences. Washington D.C.

Rees, A. R. (1965). Some factors affecting the viability of oil palm seeds in storage. *Journal of NIFOR*, **4**, 317.

Turner, P. D. (1965). Serious outbreak of blast disease in an oil palm nursery. *FAO Plant Disease Rpt.*

Turner, P. D. and Bull, R. A. (1967). *Diseases and Disorders of the Oil Palm in Malaysia*. Incorporated Society of Planters, Kuala Lumpur, Malaysia.

WAIFOR & NIFOR Farmers Booklets, Nos. 1 & 2.

Zeven, A. C. (1965). Oil palm graves in Southern Nigeria, Part 1. Types of groves in existence. *Journal of NIFOR*, **4**, 226.

Zeven, A. C. (1973). The introduction of the NIPA palm to West Africa. *Journal of NIFOR*, **5** (18), 35–36.

CHAPTER 14

The Coconut

The coconut – *Cocos nucifera* – is assumed to have its centre of origin somewhere in the Indo–Malayan region, the Latin American origin having been doubted by many authorities. Most authorities seem to favour the view that the coconut originated in the Indo–Malayan region and to the north-west of New Guinea. This is supported by the fact that over 80 species of insects have been identified with the coconut in Melanesia. It is recorded that 47 per cent of these insects live only on the coconut. The Melanesian area extends from

New Guinea to the Fiji Islands. It is very unlikely that the original home of the coconut will ever be more accurately located, the crop having spread to many parts of the world.

14.1 TAXONOMY AND BOTANY

The coconut belongs to the plant family of the Palmae, the tribe Cocoideae and to the genus *Cocos*. The specific name is *Cocos nucifera* (L.). It is a monoecious, monocotyledoneous tree with the inflorescence carrying both male and female flowers. Coconut trees are tall, stately, growing to a height of 24 m or more, generally not branching. However, dwarf coconuts also occur. The coconut stem terminates with a radiating crown of large paripinnate leaves (fronds), 6 m or more long; leaves are compound with the leaflets arranged along each side of a common rachis. The frond bears no terminal leaflet. The leaflets are many, parallel veined and very stiff. The leaf petiole is strong, broad, usually bulbous and at the base forms a clasper round nearly half the circumference of the trunk. After leaf fall, the petiolar claspers leave prominent shoehorn marks on the stem of the coconut. The phyllotaxy is two-fifths.

The roots are adventitious, fibrous and are continuously produced from the basal part of the stem. This region of the stem is usually bulbous. The roots have no cambial layer, they reach their maximum diameter quickly, function for a while and die off to be replaced by new ones.

Under good conditions of growth, the coconut palm (tall varieties) starts to flower in the sixth year of field planting. Dwarf and hybrid varieties usually flower earlier. A flower cluster (inflorescence) is produced in each leaf axil (some may abort especially if they are initiated during the dry season). An average tall variety of coconut produces 12 new fronds and inflorescences annually. The inflorescence is enclosed in a double sheath or spathe, the whole structure is the spadix. When fully grown, the spadix splits longitudinally down the underside to release the inflorescence. The inflorescence is composed of a main axis and side branches, generally 10 to 45. The female flowers are situated at the base of the main axis or on up to five branches. Each branch carries up to 200 or more male flowers. Each flower has six perianth segments in two alternating whorls, six stamens, aborted pistil with three apical teeth each bearing a nectar gland which attracts pollinating agents.

Each female flower consists of six perianth segments, thick and tightly folded over the pistil when young. Below these imbricate lobes are two bracteoles at the junction of the short peduncle. The remains of the abortive stamens lie between the ring of the perianth lobes and the fleshy ovary. The stigma is triangular in shape, extending as three erect teeth when receptive. The perianth lobes persist at the base of the fruit till maturity.

The coconut is predominantly cross-pollinated. On the same inflorescence, male flowers open about two weeks ahead of the female flowers, discharging their pollen grains before the female flowers become receptive. This ensures cross-pollination. Nevertheless, pollination from succeeding inflorescences can take place. In dwarf and hybrid varieties, male and female flowers often open simultaneously, thus allowing self-pollination to take place.

The main agents of pollination are wind and insects. Pollen grains are generally wind borne over short distances while insects, especially bees, can carry them over long distances. In addition to bees, wasps, beetles, ants and flies have been shown to effect pollination in the coconut.

Fruit maturation takes about 12 to 13 months. Immature fruit drop is common in the coconut palm, at times reaching a high level of 65 to 70 per cent.

Fruit development commences with rapid growth of the husk followed by enlargement of the cavity containing the coconut water, the development of the endosperm at about five to six months after pollination. The endosperm commences development at a point opposite the stalk and gradually extends all round the interior surface of the shell. Initially the endosperm is thin and jelly-like, later gradually becoming thicker and solid. The embryo is located near one of the three eyes and develops simultaneously with the endosperm. Towards the end of nut maturation, the weight decreases slightly owing to loss of water. Ripe nuts can be identified by shaking. On shaking the water sounds and sloshes inside the nut.

The coconut palm grows predominantly on sea shores and coastlines. As a result of this, ocean currents were recorded to be the first agent of disseminating the coconut palm to different parts of the tropical and subtropical world.

Another agency of dissemination is man. All inland coconut trees owe their existence to man. Man has also contributed substantially to the improvement of coastline coconut groves.

The major coconut producing countries of the world are the Malayan Archipelago, South-East Asian countries, India, Sri-Lanka, Pacific Territories, East Africa, West Africa, Central and South American countries.

There are two main types of coconut palm: the tall varieties, which are most common, and the dwarf varieties.

14.2 THE ENVIRONMENT OF THE COCONUT PALM

The coconut palm is at home in the warm, wet tropical climates. It thrives luxuriantly under high temperatures 27 °C–35 °C with very little diurnal variation. The crop does not give economic production in areas with long periods of drought. Good production is obtained in areas with a minimum annual rainfall of 1250 mm, evenly distributed throughout the year.

Coconut palms require plenty of sunshine. Growth in areas with a continuous cloudy sky is always poor. Strong wind is harmful to the coconut palm although the unthrifty coast line plants do withstand some degree of tidal winds. Commercial coconut plantations should be protected with a windbreak. The coconut palm grows best in the tropical lowlands. It also gives economic production in lowland subtropical parts of the world. Such areas must be protected from cold winds. As a result the coconut palm grows from sea level to an altitude of 900 m generally, but near the equator it can be grown up to an altitude of 1200 m. Although an adventitious fibrous rooted crop plant, the coconut palm grows best on well-drained, deep and fertile soil with a considerable amount of organic matter. In some coastal areas, the coconut palm grows in more or less pure sand underlain with moist infertile soils. Such coconut groves are generally unthrifty. The coconut palm is tolerant of a wide range of soil pH.

14.3 AGRONOMY OF THE COCONUT PALM

14.3.1 Selection of a Site

This follows the general principles for the selection of sites for planting tree crops. The site for coconut should preferably have sandy or light loamy soils free from waterlogging, level or only gently undulating, and protected from strong and/or desiccating winds.

14.3.2 Propagation

The coconut palm is propagated exclusively by seed. Seeds should be selected from parents which are stout, have a good and even growth and a straight trunk with closely spaced leaf scars. The number of leaves or fronds should be many and the leaves should be well oriented in a dense crown. It is essential that bunch stalks are short and capable of supporting the weight of a good number of fruits. Finally, mother trees should have inflorescences with high numbers of female flowers. Seeds should be selected from old trees of known performance, preferably trees with individual yield records over 10 or more years.

Although the coconut seed can be planted at stake in the field, the practice does not afford planters an opportunity to select outstanding seedlings for planting. This practice is, therefore, discouraged and the standard practice is to pass the seeds through a nursery stage.

14.3.2.1 Seedbeds

Coconut seeds are planted in seedbeds after the seed and the beds have been adequately prepared. Seedbeds for coconut seeds are usually narrow, long and deep. Nuts are placed in rows 20 to 30 cm apart with 20 cm between the rows.

14.3.2.2 Pretreatment of the seeds

Nuts intended for planting should be harvested when they are perfectly mature (about 12 months after pollination), or preferably such nuts should be allowed to mature on the parent trees and drop of their own accord. Freshly harvested nuts should be stored for three to four weeks before sowing.

The purpose of the short storage period is to allow the coconut embryo time to complete its dormancy period. After storage, the nuts could be given any or both of the following treatments to aid early germination:

1. The nuts can be soaked in water for one to two weeks before sowing.
2. A bit of the exocarp and mesocarp can be cut from the bottom end. The cut should not be deep enough to damage the shell.

14.3.2.3 The nursery

After the nuts have been pretreated, they are sown in the nursery beds. Nuts are placed horizontally in the planting holes, embedded in the soil but not completely buried (see Figure 40). The beds are to be watered regularly. It takes

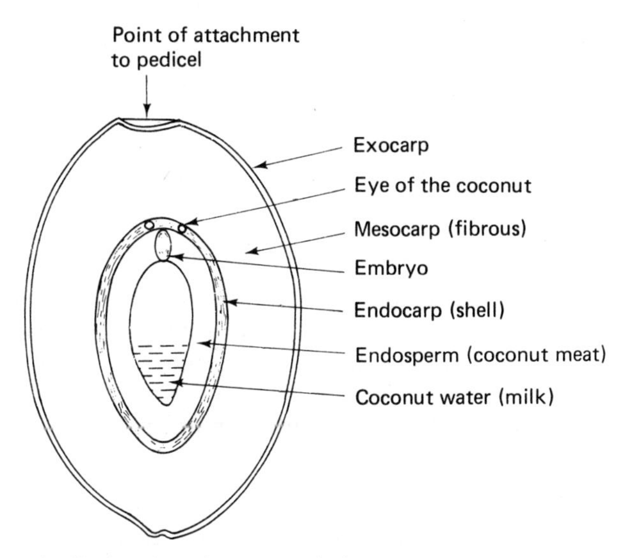

(a) Longitudinal section of a coconut fruit

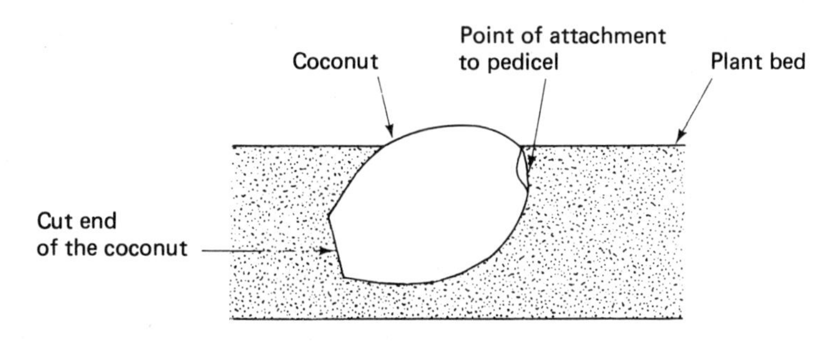

(b) Placement of the coconut in plant bed

Figure 40. The coconut and its placement in plant bed

8 to 10 weeks for fully mature nuts to germinate. When in the nursery, the seedlings should be protected from rodent attacks by fencing the nursery with wire netting; disease and pest control should be arranged when required. Weeding and application of organic fertilizers should be carried out. Application of potash, 6 to 10 weeks before transplanting into the field, has been found to be beneficial. Under normal conditions of growth, seedlings are usually ready for transplanting into the field at the age of 25 to 30 weeks in the nursery. Seedlings are at the three or four leaf stage at this age.

14.3.2.3a Germination

The coconut embryo is situated under the soft eye embedded in the endosperm. The embryo commences germination by the elongation of the plumule towards the soft eye. The lower part of the cotyledon develops into the haustorium (the coconut apple), a spongy absorbent organ, which grows slowly to fill the central cavity. As the haustorium enlarges, adventitious roots arise at the base of the plumule. Both the plumule and the adventitious roots push their way through the soft eye while the haustorium absorbs and digests nutrients from the endosperm to feed the plumule and the root system.

14.3.3 Transplanting

Field preparation requires clear-felling, after which the field is lined out. The size of the planting hole for coconut seedlings is usually 60 × 60 × 60 cm and the spacing is 9 m triangular, this giving 140 coconut palms per hectare. Seedlings are carefully dug out of the nursery with the nuts attached and placed in the planting holes at the same orientation as they were in the nursery. The soil is consolidated around the seedling.

14.3.4 Post-planting Field Maintenance

Early post-planting maintenance operations in coconut plantations include fencing of individual trees with wire netting to protect them from attack by rodents, especially grasscutters, during the first few years after establishment. At the same time a leguminous ground cover should be established. Any young plants that do not survive should, of course, be replaced with new seedlings. During later stages some palms may succumb to insect pests or diseases or they may prove to be poor growers and producers. Such palms should be replaced by new, more promising and healthy materials.

All through the productive life of the coconut palm weeds (and cover crops!) should be kept away in a circle around the trunk with a radius of 2 m. It is very beneficial to the palm when this area is kept mulched, particularly with coconut husks, otherwise discarded, and dropped palm fronds. The layer of mulch should not become too thick as otherwise the rotting vegetation may attract rhinoceros beetles as a breeding site.

Coconut palms benefit from the application of both organic and inorganic fertilizers. The recommended mixtures of inorganic fertilizer are:

Sulphate of ammonia	2 parts
Simple superphosphate	2 parts
Muriate of potash	3 parts

This mixture is applied at the rate of 0.5 to 2 kg per tree per annum depending on the age of the tree and the fertility status of the soil.

14.3.5 Major Diseases of Coconut in West Africa

The coconut palm has a low number of pathogenic diseases. The major diseases of the coconut palm are as follows:

1. *Bud-rot* A disease condition in which the growing point of the conconut palm tree becomes rotten. It is a disease associated with poor conditions of growth and excessive humidity. It is caused by *Phytophthora palmivora*, the same fungus that causes the *Phytophthora* pod rot of cacao. Once *P. Palmivora* has established itself on the growing points of the trees, other saprophytic fungi invade it also. The palm has an unthrifty apical growth, gradually leading to a complete stoppage of growth, followed by necrosis of the apical leaves and rotting of the growing point. Affected trees should be uprooted and removed from the farm or burnt.

2. *Red ring disease* A disease caused by the eelworm – *Rhadinaphelenchus cocophilus*. The eelworm invades the roots and renders them physiologically inactive. The most conspicuous symptom of the disease is the development of red rings on the unhealthy leaves. The best approach is to report any outbreak of red ring disease to the nematologists and to treat the soil around the trees with any available nematicide.

3. *Kamicope disease* This disease is probably one of the lethal coconut disease caused by a mycoplasm. Elsewhere, control of such diseases has been obtained by the use of resistant dwarf varieties in the breeding programmes.

Common *abnormalities* may be conceived as diseases. For example, normally the coconut palm has one stem, but due to damage to the growing point, branching has been recorded although very rarely. Another example is the occurrence of multiple shoots which may arise as a result of polyembryony. A nut on germination then gives rise to two or more shoots, all emerging through the same eye.

Under poor growth conditions, coconut palms may produce suckers. These provide a rare means of vegetative propagation! Suckers have been detached and grown to normal mature trees. Plant physiologists and breeders are investigating the possibility of inducing suckering in the coconut palm as a means of vegetative propagation, but have so far been unsuccessful.

4. *Button and immature nut fall* Button fall relates to nuts that fall within two months from pollination. Immature nut fall concerns larger nuts that fall before maturation. The fungus *Phytophthora palmivora* has been accused of being responsible for immature nut fall. Also, insect-induced button or nut fall is known. Both buttons and nuts may also fall due to natural causes when they are too many for the tree to provide them with sufficient water and nutrients for normal development.

5. *Tapering* This is a gradual reduction in leaf size and the size of the crown followed by yellowing of the leaf and death of trees. It has been associated with complex nutrient deficiencies, especially potassium.

6. *Nutrient deficiencies* These are well known to occur in coconut palm, especially when shortages occur in phosphorus, nitrogen, boron, iron, manganese and magnesium.

14.3.6 Major Pests of Coconut in West Africa

The coconut palm suffers damage from many pests including insects, mammals and birds. The major pests are listed.

14.3.6.1 Insects

The major **beetles** that attack the coconut are the coconut Black Beetle (*Oryctes monoceros* or the Rhinoceros Beetle), the Hope Beetle (*Xylotrupes gideon* and other *Xylotrupes* spp.) and various species of leaf-eating beetles. The beetles are very difficult to control. The major means of control are: farm sanitation (removal of breeding sites); chemical control with insecticides; and biological control through the use of predators or parasites.

The main **weevil** pest of the coconut is the Red Weevil (*Rhynchophorus ferrugineus*), the larvae of which bore into the stem of the coconut palm. Control, as with the beetles, is difficult. Attempts are being made to employ systemic insecticides for its control.

Although serious attacks by **wood borers** are not common in the coconut, a few cases of epidemics of wood borers – *Mellitoma insulare* – have been reported. Control is by uprooting of affected trees, which are to be removed from the plantation and burnt. Systemic insecticides could be used for control during the early stages of attack.

Several species of **butterflies and moths** are pests of the coconut palm. However, they never become serious pests on farms where adequate sanitation is maintained. Control is through farm sanitation and in serious cases by chemical control.

14.3.6.2 Mammals

Monkeys are a major pest of the coconut palm. It is advisable not to grow coconut palms in areas with large populations of monkeys or, alternatively, to eradicate such monkey populations before the cultivation of coconut palms is embarked upon.

Wild pigs are less harmful to coconuts than the monkeys but they may seriously affect the roots and fallen fruits.

Porcupines may attack the roots and stems, especially of young trees. They can be controlled through trapping.

Rats and flying foxes feed on immature fruits of coconuts. Control is by trapping, shooting and use of rodenticides.

14.3.6.3 Birds

Weaver Birds (*Quelea quelea*) form a very serious pest of the palms (coconut and oil palms). They nest in the crown and defoliate the trees. They can be scared away by repeated shooting and the use of avicides.

14.4 HARVESTING AND PROCESSING

14.4.1 Harvesting

Young coconut palms come to bearing from the fifth or sixth year after field planting and they can remain in production for over 60 years. Coconut fruits which are in bunches are harvested with the harvesting hook, unless the nuts are allowed to mature fully on the trees and drop naturally. Fallen nuts are then picked regularly.

14.4.2 Types of Produce Obtainable

The main product obtainable from the coconut palm is the nut. Several types of produce are obtainable from the coconut fruit. Some of these are as follows:

1. The endosperm of the fresh nut used as such or after expressing 'coconut milk' in preparation of foods.
2. Copra, the dried endosperm for oil extraction after which the coconut cake remains. Sometimes copra is desiccated and grated for use in the confectionery industry.
3. Coir, the mesocarp fibres, which can be used for weaving and plaiting. During their extraction coir dust is produced.
4. Coconut shell, which can be processed in various utensils or into charcoal.
5. Coconut fruit water.

14.4.2.1 Copra

Copra is the dried endosperm of the coconut fruit. In preparing copra, mature fruits are cut into halves and either dried naturally in the sun or in kilns. Copra is generally dried to 6 per cent water content for storage. Storage of copra at higher moisture contents encourages mould growth, bacterial and insect infestation and development of free fatty acids, thereby lowering the quality of the product.

Copra is graded for marketing. The internationally accepted grades are:

Grade I Only 'halves', well seasoned, mature, from ungerminated nuts, well dried, and of golden colour, free from dirt, mould, insect infestation, stainless and without burn marks.
Grade II Well dried with minimum of stains and burn marks. Some degree of discoloration is tolerated. All other qualities remain as in Grade I.
Grade III Badly discoloured halves, soft leathery halves, thin, shrivelled. May originate from immature or from germinated nuts.

The oil content on dry matter basis of top quality copra ranges from 66 to 70 per cent.

14.4.2.2 Coconut oil

Coconut oil, like oil palm kernel oil, is an endosperm oil. It can be extracted in a number of ways.

Crude oil or hard oil is obtained by extracting the endosperm, which is grated. The gratings are boiled in water and the oil screened off. The screened oil is refined by frying to eliminate any residual moisture.

In modern methods of processing coconut oil is extracted from copra by mechanical or solvent extraction techniques. Mechanical processes make use of screw presses or hydraulic presses. In solvent extraction the copra, after milling, is passed repeatedly through a solvent in which the oil is dissolved. Distillation of the solvent allows the oil to be obtained free of any admixtures.

The following precautions should be taken in processing copra to obtain high quality coconut oil.

1. Well dried (6 per cent moisture content) high quality copra, completely free from mould, insect and bacterial infestation must be used.
2. Copra must be quickly processed on arrival at the mill and the oil collected in clean containers.
3. 'Foots' (cake residue, mucilage, etc.) should be separated from the crude oil as quickly as possible by filtering or settling.
4. Settling tanks should be regularly cleaned.
5. The freshly extracted oil should be neutralized, deodorized, filtered and sterilized (heat to 110 °C to 120 °C) to remove any last traces of moisture as well as killing the microorganisms therein and inactivating the lipolytic enzymes.
6. The refined oil should be stored away from light and air in large drums, tankers or soldered kerosene tins especially when small quantities are handled.

Table 41 Approximate composition of coconut cake in per cent

	Hydraulic press	Expeller	Chekku
Moisture	11.00	10.00	13.30
Oil	6.00	10.00	26.70
Protein (N × 6.25)	19.80	19.10	14.30
Carbohydrate	45.30	43.80	32.80
Fibre	12.20	11.80	8.90
Ash	5.70	5.30	4.00

14.4.2.3 Coconut cake

Coconut cake is the residual cake obtained after oil has been extracted from copra. The yield averages 33 to 42 per cent from good quality copra. The main use of coconut cake is as a component of livestock feed. The composition of coconut cake varies according to the method of oil extraction. This is illustrated by the approximate composition data (Table 41) from three methods of milling. The main variation in the composition is in the oil content.

14.4.2.4 Coir fibre

Coir fibre is the fibre obtained from the mesocarp (husk) of the coconut fruit. Coir fibre has been used by man for quite a long time. It was widely used by the Arabs as early as the eleventh century. Cables and rigs made from coir fibre were used by early voyagers. Coir fibre has been manufactured into various fabrics – mats, matresses, matting, ropes, cordages and carpets. Best quality coir fibre is obtained from somewhat immature coconut fruit (maturity from the point of view of copra manufacture).

Coir fibre is processed as follows:

1. The coconut fruit is husked.
2. The husk is retted in saline water for up to 10 months, the saline water being repeatedly changed to obtain good quality fibre. Near sea shores, retting chambers can be constructed on the shore to ensure a continuous flow of saline water.
3. After retting, the husks are beaten with wooden mallets to separate the fibre. The fibre is then cleanly washed, dried and graded. Roughly 1000 coconut fruit husks yield about 80 kg of coir fibre.

In preparing bristle coir fibre, the husks are retted for only a few weeks. They are then crushed in iron rollers, and the fibres separated mechanically. The fibres are washed, dried and hackled (combed).

Coir fibre is graded for marketing. There are three international grades:

1. Three tie – individual fibres 25 to 30 cm long.
2. Two tie – individual fibres 20 to 25 cm long.
3. Soft fibre – this is of inferior quality, known as 'O' mat and sold as superior mattress fibre.

In pracitce, milled fibre gives approximately 87 per cent of its weight as bristle (three and two tie), 10 per cent of 'O' mat and 3 per cent is lost through separation.

Coir fibre is highly lignified, containing low cellulose content and is superior to other vegetable fibres for making cables as it is light, elastic, water and saline resistant as well as resistant to mechanical wear and tear.

14.4.2.5 Coir dust

Coir dust results as a by-product from the manufacture of coir fibre. Coir dust is popularly used as packing material, in the manufacture of fibre boards and of insulating material.

14.4.2.6 Desiccated coconut

This is a delicacy which is prepared from the copra after any brown parings have been scraped off. The white meat is shredded and dried at a temperature of 60 °C to 75 °C in hot air driers to a moisture content of less that 2 per cent. It is then hermetically packed. It is a human food delicacy.

14.4.2.7 Toddy

This is the wine obtained by tapping the developing inflorescence of the coconut palm. In its fresh state, toddy contains sucrose as its main constituent. Toddy can be fermented and distilled to produce alcohol or vinegar.

14.4.2.8 Coconut shell

This is the endocarp of the coconut fruit. It is used as firewood and the resultant ash is high in potash (30 to 52 per cent). It is also used in the preparation of 'gas absorbent carbons'. It can be processed into a high quality charcoal which is used for chemical filters. Finely ground, it produces the coconut shell flour which is used industrially as 'fillers in plastics' where it gives lustre to the moulded article and improves resistance to moisture. Coconut shell is used for carving smoking pipes, drinking bowls, rubber latex collecting cups, scoops and ladles, ash trays, flower vases and trinket boxes.

14.4.2.9 Coconut water

This is the liquid which is contained in the endosperm cavity. The liquid fills the endosperm cavity in immature coconut fruits while it shrinks progressively as the fruits mature. The main contents of the coconut water are water, sugars, enzymes, and vitamins especially ascorbic acid. The approximate vitamin content of coconut water is as follows:

Ascorbic acid	0.70–3.70 mg/100 ml
Nicotinic acid	0.64–0.70 mg/100 ml
Pantothenic acid	0.52–0.55 μg/100 ml
Biotin	0.02–0.025 μg/100 ml
Riboflavin	0.01 μg/100 ml
Folic acid	0.003 μg/100 ml.

Fresh, the coconut water is a pleasant drink especially when blended with gin in equal parts and chilled.

14.4.3 Coconut Fronds, Wood and Roots

The leaf of the coconut, or the coconut frond, is used for thatching, screening and for the construction of temporary walls. The midribs of the leaflets are used as brooms. The wood is used for firewood, fencing and temporary buildings. The roots are used by natives as components of medicinal preparations for the treatment of dysentery, mouth washes and gargles, and as chewing sticks.

ADDITIONAL READING

Child, R. (1953). *The Coconut, New Biology* **15**, 25–42. Penguin Books.

Child, R. (1964). *Coconuts*, Longmans, Green & Co. Ltd., London.

Cook, O. F. (1910a). The origin and distribution of the coconut palm. *US National Herbarium*, **7**, 257–93.

Cook, O. F. (1910b). History of the coconut palm in America. *US National Herbarium*, **14**(2), 217–342.

Guillaume, M. (1958). Economic rurale de l'Oceanie Française. *Agron. Tropicale*, **13**, 279–300.

Martyn, E. B. (1955). Diseases of coconuts. *Tropical Agriculture*, **32**, 162–169.

Patel, J. S. and Anandan, A. D., (1936). Rainfall and yield in coconut. *Madras Agric. J.*, **24**, 5–15.

Pieris, W. V. D. (1945). Regeneration of coconut palms. *Coconut Research Scheme* (*Ceylon*), Bull. No. 5, 9–12.

Salgado, M. L. M. (1941). The potash content of coconut husks and husk ash. *Tropical Agriculturist* (*Ceylon*), **97**, 68–73.

Tammes, P. L. M. (1955). Review of coconut selection in Indonesia. *Euphytica*, **4**, 17–24.

CHAPTER 15

Mango

Mango is the most ancient of fruits of India, where its use is intimately interwoven with Hindu mythology and rituals. Mango cultivation by the Indians probably is as old as civilization on the Indian subcontinent; the name, apparently, indicates that the mango fruit is 'the fruit of the masses'. All parts of the mango tree, from root to leaf, serve some purpose to mankind.

15.1 THE ORIGIN OF MANGO

The centre of origin of the mango (= *Mangifera indica* L.) is in the Indo–Burma region, whence it spread into other South-East Asian areas and, finally, all over the tropics. More than 60 related species, belonging to the same genus, occur in South-East Asia, with the majority in the Malay Peninsula. Fifteen of these also bear edible fruits although their quality is not comparable to that of the mango.

15.2 BOTANY OF MANGO

The mango belongs to the family of the Anacardiaceae, to which also the cashew and pistachio nuts belong. All *Mangifera* species are arborescent. Leaves are alternate, petiolate, entire and coriaceous. The mango tree is of medium size and has a deeply penetrating taproot system.

The inflorescences are axillary or terminal panicles with small and regular flowers. The colour of the flowers ranges from red to pink to white. The flowers may be hermaphrodite or male, borne within the same panicle. They are subsessile, rarely pedicellate. The calyx usually has five sepals, but this may vary from three to seven. The corolla is alternate and pentamerous, although the number of petals may vary from three to seven. The androecium consists of staminodes and fertile stamens, totalling five in number. One, or sometimes two are fertile stamens. The fertile stamens are longer than the staminodes and bear two-celled anthers, pink in colour.

The ovary is sessile and has one loculus. It is oblique and slightly compressed and situated on a disc. The solitary ovule is anatropous and pendulous. The style arises from the edge of the ovary and ends in a simple stigma. Sometimes, several carpels develop in a flower.

The mango fruit is a drupe. The butter-like mesocarp of the fruit constitutes the main edible part, although the stone (the seed) may also be consumed in some parts of the world. It is noteworthy that polyembryony of the seed is common.

The basic chromosome number of *M. indica* is 20 (i.e. $2n = 40$).

15.3 ECOLOGY

Mango can adapt itself to a wide range of climates. It thrives both in tropical and subtropical areas. Deep alluvial soils which are well drained and rich in organic matter are best for its cultivation. The lowest temperatures tolerated by the mango are 1 °C to 2 °C. On the other hand, it can withstand very high temperatures.

An important feature of the mango is its ability to perform well even in low rainfall regimes. A good crop can even be obtained with an annual rainfall of 750 mm p.a., if this is well distributed over eight to nine months. Such dry areas also fulfil the condition of a high insolation. Regularly overcast skies prevent a satisfactory development of the crop.

Strong winds can do much harm to mango trees as its wood is soft. When storms are frequent, serious damage may be inflicted.

Mango may grow up to 1200 m altitude in the tropics, but best performance can be expected in the tropical lowlands below altitudes of 600 m.

15.4 AGRONOMY OF MANGO

15.4.1 Propagation

Mango can be propagated both through seed and vegetatively. Propagation through the seed is easiest and cheapest. The seed, the mango stone, has a limited period of viability ranging from 80 to 100 days when stored under cool conditions. This is commonly done by placing the seeds between layers of charcoal or sawdust. Viability can be checked by flotation of the seed. Viable seeds sink in water. Any floating seeds should be discarded. Seed for planting should be collected from fully mature, naturally dropped fruits. Such fruits are to be collected de-pulped and air dried in a cool place. The dried seeds are planted immediately or stored for up to 80 days. Seeds are planted in seedbeds, baskets, or polythene pots. When planted in seedbeds they should be spaced at 40 to 45 cm.

In the case of polyembryonic seed the dominant seedling, apparently, often originates adventitiously from nucellar tissue. It therefore, perpetuates the maternal genotype. However, sometimes the additional seedling originates from the sexual embryo itself. Experience with the progeny will teach whether propagation through seed maintains the genotype or not. Monoembryonic seeds give rise to seedlings invariably different from the maternal genotype. Monoembryony is the rule in most Indian mango cultivars and, hence, for maintenance of a particular genotype or variety one has to resort to various methods of vegetative propagation.

Vegetative propagation of mango can be achieved by budding and grafting. It is known that budded seedlings may come into bearing within four or five years, while normal seedlings may only reach this stage after ten years. It is also possible to root mango by layering techniques.

15.4.2 Field Preparation

The standard procedure of land preparation for the establishment of mango is clear-felling, followed by burning of trash, lining out and digging of the planting holes. The holes should measure 60 × 60 × 60 cm.

The spacing used for mango varies from 9 × 9 m to 10.5 × 10.5 m depending on the variety to be planted. Dwarf varieties can be planted at 6 to 7.5 m square, while up to 12 × 12 m or even 15 × 15 m may be necessary for some varieties.

To improve the economics of commercial plantings various systems of dense planting, followed by later thinning, are recommended. As an example, the mango may be established at 6 × 6 m, which improves the early productivity of mango per hectare. Once the trees are closing in, gradually three-quarters of them should be removed in order to achieve a spacing of 12 × 12 m.

When establishing mango plantations it is advisable to provide for windbreaks as a protection against expected storms.

15.4.3 Routine Maintenance

It is advisable to establish a cover crop to protect the soil from the time of transplanting onwards. The need for this will become less when the trees are forming their canopies. Close to the trunks it is necessary to remove any weeds, including the cover crop, in order to allow the mango a fair chance to develop without undue competition.

During the first dry periods after transplanting it may be necessary to water the young plants, which by then would not have formed a sufficiently deep root system.

During the first few years flowers may have to be removed from trees that flower prematurely. It is not advisable to allow fruit formation during the first three years, because otherwise trees may be slowed down in their vegetative development. In fact, this development should be supported by the application of manure during the first four or five years. Later on, it may be advisable to provide fertilizer, particularly nitrogenous and potassium fertilizers.

As with a number of other tree crops which are spaced widely at planting, mango is usually intercropped for the first three to four years after its establishment in the field, before the canopy is closed. The intercrops are often vegetables annual and/or biennial food crops. Adequate manuring and application of fertilizers must be ensured during and after intercropping.

15.4.4 Fruiting and Harvesting

Mango is prolific in flowering but only one-third of the flowers produce fruit. Flower abortion is very high. Low percentage set of the hermaphrodite flowers may be due to frequent failure of pollen to germinate after deposition on the stigma, failure of the proper development of the gynaecium or thrips damage. Flower abortion may also be related to abnormal pairing of chromosomes at meiosis. This should be further investigated.

Mango flowers during the months of January and February, in West Africa. The fruits mature in April and May. Once mature, the fruit is easily and rapidly perishable. Upon maturation, the fruits should be immediately made available for consumption or, alternatively, be preserved.

Mango fruits are harvested by hand and collected in bags. The harvester with a collecting bag climbs the tree, harvests the fruits one by one into the bag, descends, discharges the content of the bag and repeats the operation until the ripe fruits on each tree have all been harvested. During the harvest season, each tree is visited weekly.

One of the problems requiring attention in mango is irregularity of fruit ripening. This phenomenon has made the harvesting of mango very expensive.

15.5 UTILIZATION AND PREPARATION OF MANGO FRUITS

Mango fruits are difficult to keep for any length of time and should, normally, be consumed within a few days from full maturity. The fruits are attractive due to their texture, flavour and colour, in addition to the highly appreciated mango taste. The fruit has the following main constituents in percentage by weight: seed or stone 12.5 per cent; skin 10.8 per cent; pulp 76.7 per cent.

Because of the high perishability of the fruits several means of preservation have been developed:

1. *Freezing* Fruits can be preserved for 30 to 40 days when frozen immediately upon harvesting. Nutritional or taste qualities remain the same throughout this period.

2. *Canning* Mango fruits may be canned as peeled fruits, slices or in the form of extracted juice.

3. *Drying* This is not extensively practised and then only in (semi) arid areas where other outlets are not available.

Also the mango seed or stone can be utilized. It is composed of the endocarp, the testa and the embryo with its large cotyledons. The endocarp and the testa are waste products, although the former can be used as fuel if available in large enough quantities. The cotyledons may be used for several purposes:

1. Boiled kernels are eaten in India in times of food scarcity.
2. Baked kernels are consumed both in India and the Sahel zone of West Africa.
3. Kernel flour is obtained by washing crushed mango kernels until all tannins are removed. Subsequently, after drying, a flour is prepared which can be served as mango porridge, reputed to be a delicacy.
4. Starch can be obtained by soaking kernels, for six to eight hours, in water to which sulphur dioxide has been added. The material is then crushed, pounded and strained. By further processing in water containing sulphur dioxide a very pure starch can be procured.

15.6 PESTS OF MANGO

The mango tree is relatively pest free. Among the most important of the few pests which attack the mango are the following:

1. *Mango hoppers* These are jassids of the genus *Idiocerus*. These pests usually hide in cracks in the bark of the mango tree. At the time of oviposition, they oviposit large numbers of eggs within the florets, causing physical injury and, subsequently, dropping of florets. The jassids also suck the juice from the inflorescence, which results in wilting and eventual death of flowers.

Jassids can be controlled by spraying with appropriate insecticides. However, the size of the mango tree often makes this impossible. Because of this attention has been given to biological control with the aid of hopper parasites (*Pipimculus*, *Pyrilloxenas* and *Epipyrops* species).

2. *Giant mango mealybug* The most common and important bug of the mango tree is the *Drosicha stebbingi* of the family Coccidea. This bug feeds on succulent parts of the plant and it is easily controlled by spraying with appropriate insecticides.

Other problems are sometimes caused by scale insects, which can be controlled with a light oil spray and by termites, which can be controlled by banding the trunk with an insecticide.

15.7 DISEASES OF MANGO

Fruit anthracnose is the most common, widespread and dangerous disease of the mango fruit. It is caused by the attack of the fungus *Colletotrichum gloeospoides*. The fungus attacks mango fruits at any stage of development, causing stunting and eventual drop of the attacked fruits. Fruit anthracnose can be controlled by spraying a fungicide. Once the fungus has been observed in an orchard, it is advisable to spray the orchard as a routine at the start of every fruiting season.

A second important disease is mango scab, which may attack any part of the tree, i.e. leaves, twigs, flowers, fruits. It is caused by *Elsinoe mangiferae*, a fungus. Control can be achieved through farm hygiene and by spraying with appropriate fungicides.

15.8 REHABILITATION OF MANGO PLANTATIONS

Rehabilitation of a mango plantation may become necessary when the trees become old or when they have been devastated by external agencies such as storms, wild animals or diseases. Several methods of rehabilitation may be followed:

1. Coppicing of the old or damaged trees. This can be done gradually so as to guarantee some harvest during the period covered by rehabilitation.
2. The old and/or damaged trees are interplanted with young seedlings (budded or otherwise) while the old trees are progressively removed as the young ones develop.
3. The use of an appropriate vegetative propagation method, such as top-working or approach grafting.

Rehabilitation necessitated by disease infections is carried out by uprooting of the infected trees and thorough clearing of the debris of the infected plants. New material is then established.

It is essential to apply appropriate fertilizers and/or manure to rehabilitated plantations.

ADDITIONAL READING

Lal Behari Singh (1960). *The Mango: Botany, Cultivation and Utilization.* World Crop Books. Leonard Hill (Books) Ltd, London; Interscience Publishers Inc., New York.

Gunaratham, S. C. (1946). The cultivation of mango in the dry zone of Ceylon. *Tropical Agriculture*, **102**, 23.

Mallik, P. C. and B. N. De (1953). Manures and manuring of mango and economics of mango culture. *Indian Journal of Agriculture Science*, **22**, 151.

Singh, L. B. (1969). Mango. In F. P. Ferwerda and F. Wit (eds). *Outlines of Perennial Crop Breeding in the Tropics.* Agric. Univ. Wageningen, Misc. Paper 4.

CHAPTER 16

Tea

Tea has its origins in South-East Asia, from whence it spread into China where the leaves were first used as a beverage several thousand years ago. The largest tea producers are India and Sri-Lanka, followed by many East and South-East Asian countries, several East African countries, and Brazil, Peru and the Argentine in South America.

16.1 BOTANY OF THE TEA PLANT

Camellia sinensis (L.) O. Kuntze is a plant species which belongs to the family of Theaccae, which encompasses some 200 woody plant species in the warmer regions of Asia and South America. Within the species, one may distinguish the China teas, slow growing dwarf trees with a good tolerance of cold weather and other adverse conditions and Assam teas, which are faster growing teas adapted to warmer conditions. The tea plant is mainly cross-pollinated and propagation has, for long, been through seed. Hence a wide range of characteristics, originating from the two types of teas, may be present in any planting of tea.

Tea is a small evergreen tree, up to 10 m high at full maturity. Under cultivation, it is restricted to low spreading bushes through pruning techniques. The plant forms a strong taproot. Leaves are alternate, exstipulate, evergreen, obovate-lanceolate, leathery glossy, serrate and may vary from 3 to 10 cm or even more. Leaves are produced in flushes from a bud in the axil of a terminal leaf of a shoot, which has attained full size. The bud usually produces two scale leaves, then a small blunt leaf, followed by four normal leaves. After elongation of the internodes, the shoot becomes dormant and a new bud develops.

The flowers are pedicellate, placed axillary, solitary or in clusters of two to four. The calyx is persistent, there are five separate petals and numerous stamens of which the filaments each carry two anther lobes. The stigmatic surface of the pistil is pentafid. The plants are largely self-incompatible and cross-fertilization has led to a high level of heterozygosity in tea. Fruits are brownish green three lobed capsules opening at maturity (after 9 to 12 months) by splitting into three valves. Each of the three locules contains one or two globose or flattened seeds. These have to be planted immediately after ripening as viability is retained for a short time only.

16.2 ECOLOGY OF THE TEA PLANT

Cultivation of the tea plant is restricted to subtropical regions and mountainous areas of the tropics, where altitudes are in the range of 1200 to 1800 m above sea level with temperature regimes of 10 °C to 27 °C without frost. This explains why tea can be expected to perform well at very few locations in West Africa. For example, only towards the eastern border of Nigeria does the Mambilla Plateau attain sufficient height to allow for tea cultivation. Further east, in Cameroun, several sites could be available, but these are outside the area presently considered. Further west the higher areas either lack the rainfall required for successful tea production or the soils are unsatisfactory; an example is the Jos Plateau in Nigeria.

Tea requires high rainfall and humidity almost throughout the year. An annual rainfall lower than 1100 mm is considered marginal. Soils for tea should be deep, well drained and may be acid.

16.3 AGRONOMY OF TEA

Areas to be planted to tea should be carefully cleared. Wood materials should be removed from the field, but not burned as patches of wood ash prevent satisfactory development of tea plants. Couch grass (*Stoloniferous digitaria*) should not be allowed on land meant for tea planting. On slopping land, contour terraces should be constructed and proper drainage ensured.

Tea is commonly propagated by seed. Tea seeds lose viability fairly rapidly and fresh seeds are, therefore, sown immediately they become available. They are sown in seed boxes and, after germination, the young plants are transferred into polythene bags. In this way the nursery period takes at least six months. The seedlings are then transferred from pots to nursery bed where they stay for a much longer period, of two to three years, before they are transplanted into the field in the form of 'bare-root' stumps.

At planting time the potted seedling or the 'bare-root' stumps should be transplanted into planting holes of 30 cm wide and 30 cm deep, spaced at 100 to 120 cm.

An alternative method of obtaining plant material through vegetative propagation is becoming widespread. This is done by placing single internode cuttings, into shaded nursery beds, often covered over with a transparent polythene sheet at some height above the cuttings. It takes about 12 months for cuttings to be ready for transplanting into the field. When using this technique, it is of the greatest importance to take for rooting only the highest yielding tea bushes. This principle also applies to other methods of vegetative propagation of tea, such as budding, grafting or the use of root cuttings.

The terminal buds of young plants should be removed to encourage the development of lateral shoots. This is done to obtain a low spreading bush of which the upper surface forms the 'plucking table' at a convenient height. When this level rises too high the bush is cut back severely to a height of 30 to 40 cm to form a new plucking table.

Weed control is of importance to allow for easy harvesting of clean produce devoid of admixtures, in addition to permitting satisfactory growth of the tea bushes. Between young plants weeds are best controlled manually. Among mature bushes herbicides may be used.

As the main produce from the tea bush is formed by newly developed leaves, fertilization with nitrogen, (known to promote vegetative growth of plants), is of the greatest importance. The application of sulphate of ammonia is recognised practice in many tea growing areas. Fertilizer application rates should be based on the results of soil analysis.

When dry periods are experienced, production drops and when drought persists, the young tips of the branchlets may die back. Production levels can be kept at high levels if irrigation is used. If water is available cheaply and in satisfactory quantities, irrigation can be very profitable.

Pests and diseases are not usually of great importance in tea plantations. An occasional attack by leaf eating caterpillars or various fungal leaf spots can be kept under control by spraying. In this respect, it is important to guard against the application of chemicals during or shortly before harvesting tea leaves.

16.4 HARVESTING TEA

The harvesting procedure for tea consists of the removal of young vegetative shoots, preferably in the stage that they show an opened leaf and a bud. The youngest tips give rise to high quality tea. The harvest has to be repeated regularly every 10 to 14 days, depending on the growth of the tea bushes. An important aspect of the harvesting of tea is the need to ensure an immediate transportation of the produce to the processing factory.

16.5 THE TEA OF COMMERCE

The main commercial product is black tea which is prepared by fermentation processes from the young leaves of the plant. The different types of tea are dependent on the method of processing, curing and fermentation. During the processing caffeine is liberated in association with tannins. The caffeine content of processed tea leaves may vary from 2.5 to 5.0 per cent. The quality of tea depends on the method of processing, and also on the environment under which the produce was grown and harvested. Considerable differences in tea quality are caused by the temperature regimes during growth of the leaves (i.e. altitude of area of cultivation), the soil type, its pH value, and the season of the year at which the tea was harvested. Commercial tea is obtained by blending different teas together. This is done is order to meet the specific requirements of different tea markets.

Popular types of tea are:

1. *Black tea* obtained by post-harvest fermentation of withered and rolled tea leaves through factory processing.
2. *Green tea* obtained from unfermented tea leaves. This product is typical of the use of tea in China and Japan.
3. *Oolong tea* obtained from slightly fermented tea leaves, is mostly produced in Taiwan for use in North America.
4. *Brick tea* prepared from the residues obtained after the preparation of black and green teas. This product is used mainly in Tibet and other parts of Central Asia.

ADDITIONAL READING

Dougan J. *et al.* (1978). *A Study of the Changes in Black Tea During Storage*. Tropical Products Institute, London.

Eden T. (1970). *Tea*, Longman, London.

Ogutuga D. B. A. (1970). Biosynthesis of caffeine in tea callus tissue. *Biochem J.*, **117**, 715–720.

CHAPTER 17

Underexploited Tropical Tree Species

There are many tree species in the tropics which are potentially useful but which still remain unknown to the international market and are not exploited. A few of these plants will be discussed while others will only be listed in this chapter.

17.1 SYNSEPALUM DULCIFICUM (THE 'MAGIC PLANT') (IGBAYUN, YORUBA)

The 'magic plant' thrives in West African rain forest areas. It is common on farms and in secondary bush, preferring moist localities. It also occurs on the fringes of virgin forest. With progressing deforestation and more intensive agriculture, the 'magic plant' is becoming less abundant in a number of its usual habitats.

17.1.1 Botany of the 'Magic Plant'

Synsepalum dulcificum is a shrub or small tree with hard wood. It has glabrous, obovate to obovate-oblanceolate leaves which occur in clusters near the ends

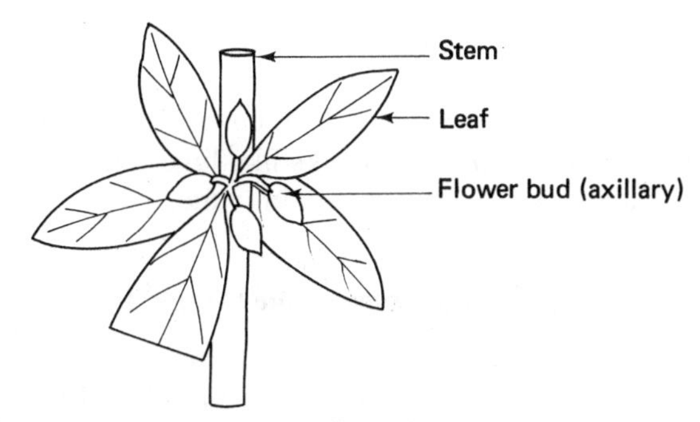

(a) Twig of 'magic plant' with flower buds

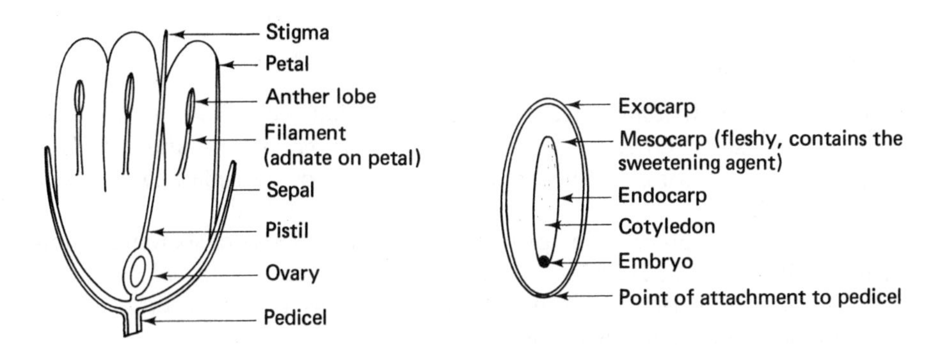

(b) Longitudinal section of the flower of the 'magic plant'

(c) Longitudinal section of the fruit (berry) of 'magic plant'

Figure 41. Floral structures of the 'Magic plant'

of branchlets. The leaves have short petioles and rounded or shortly acuminate apices. The brown flowers are subsessile in small axillary clusters; the calyx is ribbed; the corolla tube is narrow and as long as the calyx. The fruit is a one-seeded berry (Figure 41).

The berry is small and turns red when ripe. The pulp of the fruit is eaten. It contains a sweet-tasting protein known as miraculin. Miraculin is similar to molenin which occurs in *Discoreophyllum cumminsi* or thaumatin which occurs in *Thaumatococcus daniellii*. Miraculin, molenin and thaumatin are claimed to be protein based (Cagan, 1973). Their sweetness relative to sucrose as estimated by Cagan is shown in Table 42.

Table 42 Sweetness of miraculin, molenin and thaumatin relative to sucrose

	By weight	By molar
Miraculin	4500	120,000
Molenin	3000	90,000
Thaumatin	750–1600	30,000

Miraculin is widely accepted as an active sweetener. On eating the pulp of the berry, the taste buds in the mouth register a strong sweetness, which may last for up to 18 hours. This reaction is so intense that substances such as bitter drugs (quinine, nivaquine) or acid drinks like lime juice, taste sweet. The pulp of the berry is widely used to sweeten palm wine.

17.1.2 Agronomy of the 'Magic Plant'

The 'magic plant' grows best when rainfall is not less than 1200 mm p.a. and is distributed over at least eight months of the year. Little is known as yet about cultivation techniques, because the 'magic plant' is not yet planted on any scale.

17.1.2.1 Propagation

Observations at Ibadan, Nigeria, have shown that seeds from mature berries germinate readily. Growth of the seedling, however, is slow and it has to be kept in a nursery for 8 to 10 months, before it can be transplanted successfully into the field.

Also vegetative propagation proved to be possible. Twigs of six to nine months root easily in about 12 to 16 weeks. The process can be accelerated by treatment of the cuttings with various growth hormones as indole acetic acid indole butyric acid, and gibberellic acid.

Rooted cuttings can be successfully transplanted into the field three to four months after hardening and when the rains are steady. Rooted cuttings and seedlings have been planted in the field at spacings varying from 5 × 5 m to 7 × 7 m with subsequent good growth.

17.1.2.2 Management techniques

The 'magic plant' does not require shade when being transplanted into the field in planting holes of 60 × 60 × 60 cm. The seedling should, of course, be kept free of weeds. During the dry seasons, especially the first one or two, watering of the plant ensures good growth.

Responses to normal management techniques are not known. It is, therefore, of importance that the influence of pruning, and the application of fertilizers or mulch be investigated. The 'magic plant' has so far been relatively free from pests and diseases. It is likely that problems would develop when the plant is really taken into cultivation.

17.1.2.3 Harvesting

On ripening, the berries turn from green to reddish yellow. At this time the berries should be hand picked. No damage to the berry should occur. Upon washing, the berries may be packed into polythene containers. If kept in a cool place the berries may remain fresh for one to two weeks. Storage in refrigerators or deep freezers prolongs keeping time. Upon decomposition or after sun drying the sweetening properties of the berries are lost.

17.2 DIOSCOREOPHYLLUM CUMMINSI

Dioscoreophyllum cumminsi is another 'miraculous' source of a sweetening agent (molenin) that may have economic potential. This plant is a climber with rusty stems attaining great heights. The large, entire leaves are variable in shape, usually three to five lobed, sometimes caudate-sagittate. The flowers are green and have yellow anthers. They are placed in hairy racemes. The fruits occur in large clusters and are red in colour. The plant forms tubers, which are edible. Various parts of the plant are used for medicinal purposes (tubers, stems, bark, berries).

The ripe fruits or berries are eaten and have a strong sweetening effect. This may be a source of a non-sucrose sweetening agent of possible use in the manufacture of delicacies.

17.3 THAUMATOCOCCUS DANIELLII

Thaumatococcus daniellii is a third 'miraculous' plant with a sweetening agent (thaumatin) capable of development for industrial use. The plant occurs wild in the African rain forests. It has a simple, very large, caudate leaf (25 × 30 cm) borne on a petiole of 2 to 4 m long. The main stem is a rhizome which is subterranean. This is also the case with the fruits.

The flower emanating from the rhizome is carried on a short peduncle just above ground level. The calyx is green, the corolla lilac. The stigma is fleshy and trifid. The tricarpellary fruit turns from green to red when maturing. It is trilocular and contains three fleshy seeds. The pulp around each seed contains the sweetening agent, which is widely consumed.

17.4 CHRYSOPHYLLUM ALBIDUM (THE WHITE STAR APPLE)

Chrysophyllum albidum is a medium-small buttressed tree species, up to 25–30 m in height with a mature girth varying from 1.5 to 2 m. The bole is usually fluted and has a thin bark with gummy latex. The leaves are simple, elliptic to oblong, or obovate elliptic, acuminate. Flowers are small, yellow and borne on small stalks. Fruits are almost spherical, slightly pointed at the tip, orange-red sometimes with speckles. Fruits have five brown seeds in yellowish pulp. The seeds measure 1 to 1.5 by 2 cm with one sharp edge. The transverse section of the fruit shows the seeds arranged star-like.

The white star apple is fleshy and juicy. The fruits are popularly eaten. The juice of this fruit is a potential source for a soft drink. It can also be fermented for the production of wine and/or alcohol through distillation.

17.5 DACRYODES EDULIS (AFRICAN PEAR)

Dacryodes edulis is a hardy species with medium sized trees up to 25 m high, with a low spreading canopy. The bark is thin and on young branchlets rusty-puberulous. The leaves are pinnate with leaflets measuring 3 to 4 cm by 2 to 3 cm. The leaflets are glabrous, narrowly oblong-elliptic, abruptly acuminate.

The flowers are fragrant. They are placed in terminal panicles; the inner flowers are cream coloured, the outer ones golden. Sepals and petals are covered with stellate scales. The fruits are orange-red to purple and have a pleasant smell. They consist of a large seed surrounded by a thin mesocarp; the pulp is eaten.

The fruits are rarely taken raw; they are usually boiled or roasted and easten alongside a main dish of maize.

These fruits are not produced from specially planted trees. They are gathered from existing trees, which are retained when forest or secondary bush is felled.

17.6 CARAPA PROCERA (THE MONKEY KOLA)

Carapa procera, the monkey kola, thrives in the West African rain forest. It is a medium sized tree, attaining heights from 20 to 25m on maturity.

17.6.1 Botany of the Monkey Kola

Carapa procera is a tree with a straggling habit, often with more than one stem. The bark is thin, smooth, greenish grey with a pink to orange 'slash'. The leaves are pinnate, 0.75 to 1.00 m in length. They are shiny, usually terminal and wine-red when young. There are 10 to 16 pairs of leaflets, which are elongate-oblong to elliptic, entire, acuminate and glabrous. Flowers are creamy white with reddish centres, sweet scented and ground near the end of branches in slender, many flowered panicles. The fruits are about 15 cm long, ellipsoid to spherical, beaked and ribbed. They break into six to eight segments, releasing 15 to 20 seeds like the kolanut. The seed is 4 to 5 cm in diameter, obliquely avoid with two flattened surfaces. Their surface is rough and reddish brown. The kernel is oily.

17.6.2 Uses of the Monkey Kola

The wood of *C. procera* is heavy, hard and strong. It can be used for flooring and is used locally for cutting of paddles, although it has a tendency to warp. It is useful for building because of its resistance to fire and to termites, and for furniture. The twigs are used as chewing sticks.

The fruits may be dried and used for the extraction of oil, which can be used for the preparation of food.

Irvine (1961) has recorded a large number of medicinal uses; among these are the treatment of sores, burns, rheumatic pains, ringworm with the oil, which can also be used as a vermifuge. Its purgative action is like that of castor oil. Local beliefs indicate that '201 different illnesses' can be cured with products from the monkey kola tree. The potential medicinal wealth of *C. procera* is yet to be explored and, perhaps, to be exploited.

17.7 BUTYROSPERMUM PARKII (THE SHEA BUTTER TREE)

The shea butter tree is indigenous in the West African savanna zone, occurring abundantly all across West Africa up to the frontier of the Sudano–Sahelian zone.

It is a small tree up to 15m in height and has a spreading canopy. The bark is thick, deeply fissured and fire resistant. The leaves are terminal, oblong or obovate-oblong, glabrous when mature and have a wavy margin. The tree produces flowers during the dry season. The flowers are creamy white, sweet scented and are placed in terminal clusters. The fruits contain one or two seeds, which are brown, shiny and have a large hilum and thick testa.

The fruit pulp is eaten and oil can be extracted from the seed, of which the kernel contains 45 to 55 per cent oil, the shea butter. The shea butter is prepared by roasting, pounding or grinding and boiling the kernels. The oil can then be skimmed off. The shea butter is used in the manufacture of soap, candles and butter substitutes. At present, seeds are collected from naturally occurring trees whenever a shortage of oils causes the price of shea butter or shea oil to rise. The crop could be developed for regular farming in the dry savanna areas.

17.8 TETRAPLEURA TETRAPTERA (THE AIDAN TREE)

The Aidan tree is a deciduous tree species commonly growing on the fringe of the West African rain forest belt. The tree reaches 20 to 25 m in height and may attain a girth of 1.5 m. It has sharp buttresses. The leaves are sessile, bipinnate with six to eight pairs of pinnae, placed opposite to each other. The pinnae have 8 to 12 pairs of oblong-elliptic, rounded leaflets. The flowers are pinkish cream, turning orange and are placed solitary or in pairs in axillary or terminal inflorescences. There are 10 stamens of which the anthers carry a gland at the apex. The fruit is about 25 cm long with four lingitudinal, wing-like ridges nearly 3 cm broad. Two of the wings are woody and the other two are filled with soft sugary pulp, oily and aromatic. The fruits are shiny, glabrous and brownish purple. The seeds rattle in the pods, are small, black and hard. The kernel contains oil.

The fruit pulp is rich in sugars, tannin and a little saponin and may be used in flavouring of food. The fruits and flowers are used as perfumes in pomades prepared from palm oil. Other parts of the plant, namely leaves, bark, roots, kernel, are used for medicinal purposes.

The Aidan tree is another West African tree of which possible uses are to be explored.

17.9 OTHER PROMISING TREE SPECIES

Several other tree species occur in the West African forest vegetation and could be developed as crops for farming enterprises, provided their uses are further explored. Examples are as follows:

1. *Blighia sapida* The immature fruits are a source for natural detergents. The endosperm of the seed contains a useful oil. The arillus is recognized as a delicacy.

2. *Raphia hookeri* Palm wine and building materials are well known traditional products of the raffia palm. Stipular and epidermal fibres from immature leaves are a source of 'raffia' used widely in horticulture and handicraft.

3. *Spondias mombia or hogplum* The fleshy fruits of this West Indian tree are used in West Africa for food. Its leaves have medicinal uses.

4. *Tetracarpidium conophorum* The seeds are rich in oil and protein and can be used for food.

There are many other tree species which are held in stock by the West African rain forests, and which could be useful to mankind. These valuable species are, however, rapidly being depleted by economic development, increasing and rapid urbanization, intensification of agricultural enterprises, and deforestation for the construction of roads, industries, etc. International efforts are required to preserve, in appropriate economic gardens or plant museums, important and potentially useful plant species, which otherwise might be lost.

ADDITIONAL READING

Ainslie, J. R. (1952). *List of Plants used in Native Medicine in Nigeria.* University of Oxford, Imperial Forestry Institute, Paper No. 27.
Cagan, J. (1973). Protein sweeteners from tropical plants. *Phytochemistry*, **4**, 145.
Githens, T. S. (1948). *Drug Plants of Africa.* University of Pennsylvania Press, Philadelphia.
Higginbotham, J. D. (1979). Talin – A novel sweet protein from *T. Daniellii* (Zingiberaceae). *Proc. Ergob. Conf. Sugar Substitutes.* B. Guggenheim (ed.). Karger, Basel, Switzerland.
Irvine, F. R. (1961). *Woody Plants of Ghana.* Oxford University Press, London.
Miller, O. B. (1952). Woody Plants of Bechuanaland. *J. S. Afr. Bot. 18*, 1–100.

CHAPTER 18

An Outlook on the Future

Cultivation of the traditional tropical tree crops in West Africa has so far been on an extensive basis. The extensive system of production will undoubtedly change and give way to more intensive systems of production. The change, which is sequential to population growth increasing demand on land for other economic and developmental activities and increasing demand on land for the production of locally consumed staple foods, has started. For more intensive systems of production, tree crop scientists will need to produce new varieties of the tropical tree crops. The onerous duty will devolve most on the plant breeders who within a very short time will be required to develop these new varieties which in addition to possessing the traditional attributes of disease resistance or escape, pest and drought (environmental) resistance, will have to be smaller in stature, more productive per unit of land area, and more responsive to advanced cultural practices such as closer spacing and fertilizer application.

In most parts of the tropics, virgin soils have become exhausted. The soils, therefore, can no longer support the traditional tree crops without an intensive application of fertilizers. In these same areas, research work in fertilizer application to several tree crops is in its infancy. In some parts, the soils have not been surveyed and soil maps are not available. Work is yet to be done on the micro-nutrient requirements of these crops.

The need for concerted international cooperation for finding immediate solutions to these problems has arisen.

There exists a considerable body of knowledge on the use of fungicides and pesticides for some of the tropical tree crops. The use of insecticides on cacao, which had resulted in the development of resistance to the cyclodiene insecticides in the mirids, vividly demonstrates the futility of complete dependence on chemical control of diseases and pests in plantation crops. The indication is clear that future emphasis needs to be placed on **pest management** which will involve the combined use of resistance/escape, farm hygiene, biological control and a minimal use of chemicals. In this regard, attention needs to be called to the increasing cost of copper and other metal bases for the manufacture of pesticides. More investment is needed on research which is aimed at the development of completely organic synthetic fungicides and insecticides. The dwindling tropical forest abounds with many plant species that have been recorded to possess fungicidal and/or insecticidal properties. International cooperation among pesticide and fungicide scientists is needed to confirm the potential usefulness of such plants, their conservation and utilization.

Weed control has always been a labour-expensive item in the cultivation of tropical tree crops. Hand labour has been mostly used. With the ever-growing scarcity and high cost of labour coupled with uncertainties which abound around the residual effects of a continuous use of herbicides in a plantation crop, the indications are that more work is needed on the mechanization of weed control in plantation crops. Another attack on the weed problem is through an efficient use of cover crops. This aspect of the management of plantation crops needs to be improved.

The cultivation of most tropical tree crops has suffered from the dearth of research relating to cost of production, labour management and comparative economic returns when the cultivation of plantation crops which are mostly exported, is compared with the cultivation of locally consumed food crops. With the changing economic climate in the traditional tropical areas where these plantation crops are produced, the need has arisen for intensive studies to be carried out on the economics of the tropical plantation crops. Data from such studies will be most useful to planners of the economies of tropical countries and to plantation managers, attempting to improve the level of farm management.

At a slightly more technical level, a number of basic problems need to be studied in most tropical tree crops. A few of these problems are briefly outlined below:

1. *Incompatibility systems* It has been shown that incompatibility systems exist in some of the tropical tree crops. Incompatibility affects the yields of these crops by reducing fruit set and lowering the number of seeds per fruit. Research work is needed on methods of overcoming incompatibility where it occurs in these crops. Also more work is needed on how best to handle the incompatibility system in the production of hybrids without necessarily concentrating incompatibility factors.

2. *Germplasm collection* All the tropical tree crops have their centres of origin in the tropics and the subtropics. Most tropical parts of the world are becoming opened up and, with this, valuable variants of most of the tropical tree crops are lost. A new airport, a new university campus, a new 1000 km express road, a new capital city, a new oil well, a new factory for iron industry, cement and the mere expansion of cities and villages are daily irretrievably swallowing up valuable genes of these tree crops. There is an urgent need, therefore, for the exploration, collection and conservation of these disappearing but valuable genes at internationally controlled germplasm centres.

3. *Cytological and genetic studies* The breeding of most tropical tree crops – cacao, coffee, kola, rubber, the palms – has been carried out without much information on the cytogenetic systems of these plants. In these days of genetic and/or cytogenetic engineering of plants and animals, future progress in the breeding of these crops is likely to be facilitated by a better understanding of their cytogenetic system.

4. *Product development and utilization of by-products* Product development and utilization of by-products of tropical tree crops constitute two major areas that have been neglected. Food technology and biochemical research are advanced enough to develop alternative uses for cacao beans, rubber, the mango seeds, to mention a few. Also, world attention should now turn to such sweetening agents as monelin, thaumatin and miraculin which are all protein based and are several thousand times sweeter than sugar (on a molar basis).

Many by-products of tropical tree crops have hitherto not been utilized. A few of these products are cacao husks, coffee pulp, endocarp of mango. These by-products can profitably be employed by man. There are a number of publications which have indicated the possibilities for their commercial use. Finding economic uses for the by-products of tropical tree crops will undoubtedly be beneficial to both the planters and the consumers. This is an area where more investment is needed.

Again research workers on tropical tree crops need to devise a more effective means of prompt dissemination of worthwhile research results to planters. There is a considerable amount of adoptable research results in a number of tropical tree crop research institutes such as: The Cocoa Research Institute of Nigeria (CRIN), The Cocoa Research Institute of Ghana (CRIG), and The Institute Francaise du Cacao et du Cafe (IFCC). Most of these institutions depend on external agencies such as Ministries of Agriculture, commercial chemical firms, publishers of journals, for the dissemination of research results. This certainly is not good enough. The arrangement which may satisfy the needs of the immediate future will be for each research institute to have its own extension service.

It is in the interest of the whole world that tropical tree crops continue to be increasingly productive. To achieve this, however, needs the full cooperation of both the producer and the consumer. In this sense, the raw cacao bean buyers (the chocolate manufacturers) need to undergo a serious self-examination and come into price range levels that are mutually agreeable to both sides. In a case where the planters cut down tree crop plantations, the alternative is immediately to put such lands into the production of locally and internationally consumed and NEEDED staple foods. The conversion of manufacturers' long-existing processing factories for other uses might be a much slower process. Therefore, cooperation and mutual understanding are a must to ensure a bright future for the continued cultivation of the tropical tree crops.

Index